SpringerBriefs present concise summaries of cutting-edge research and practical applications across a wide spectrum of fields. Featuring compact volumes of 50 to 125 pages, the series covers a range of content from professional to academic.

Typical topics might include:

- A timely report of state-of-the art analytical techniques
- A bridge between new research results, as published in journal articles, and a contextual literature review
- A snapshot of a hot or emerging topic
- An in-depth case study or clinical example
- A presentation of core concepts that students must understand in order to make independent contributions

Briefs allow authors to present their ideas and readers to absorb them with minimal time investment. Briefs will be published as part of Springer's eBook collection, with millions of users worldwide. In addition, Briefs will be available for individual print and electronic purchase. Briefs are characterized by fast, global electronic dissemination, standard publishing contracts, easy-to-use manuscript preparation and formatting guidelines, and expedited production schedules. We aim for publication 8–12 weeks after acceptance. Both solicited and unsolicited manuscripts are considered for publication in this series.

Indexing: This series is indexed in Scopus, Ei-Compendex, and zbMATH

Hua Xu • Hanlei Zhang • Ting-En Lin

Intent Recognition for Human-Machine Interactions

 Springer

 清華大学出版社
TSINGHUA UNIVERSITY PRESS

Hua Xu 🆔
Department of Computer Science
and Technology
Tsinghua University
Beijing, China

Hanlei Zhang 🆔
Department of Computer Science
and Technology
Tsinghua University
Beijing, China

Ting-En Lin 🆔
DAMO Academy
Alibaba Group
Beijing, China

ISSN 2191-5768   ISSN 2191-5776   (electronic)
SpringerBriefs in Computer Science
ISBN 978-981-99-3884-1   ISBN 978-981-99-3885-8   (eBook)
https://doi.org/10.1007/978-981-99-3885-8

Jointly published with Tsinghua University Press
The print edition is not for sale in China (Mainland). Customers from China (Mainland) please order the
print book from: Tsinghua University Press.

This Springer imprint is published by the registered company Springer Nature Singapore Pte Ltd.
The registered company address is: 152 Beach Road, #21-01/04 Gateway East, Singapore 189721,
Singapore

Preface

The natural interaction ability between human and machine mainly involves human-machine dialogue ability, multi-modal sentiment analysis ability, human-machine cooperation ability, and so on. In order to realize the efficient dialogue ability of intelligent computer, it is necessary to make the computer own strong user intention understanding ability in the process of human-computer interaction. This is one of the key technologies to realize efficient and intelligent human-computer dialogue.

Currently, the understanding of the objects to be analyzed requires different levels of ability such as recognition, cognition, and reasoning. The current research on human-computer interaction intention understanding is still focused on the level of recognition. The research and application of human-computer natural interaction intention recognition mainly includes the following levels: intention classification, unknown intention detection, and open intention discovery. Intention understanding for natural interaction is a comprehensive research field involving the integration of natural language processing, machine learning, algorithms, human-computer interaction, and other aspects. In recent years, our research team from State Key Laboratory of Intelligent Technology and Systems, Department of Computer Science and Technology, Tsinghua University has conducted a lot of pioneering research and applied work, which have been carried out in the field of intention understanding for natural interaction, especially in the field of intention classification, unknown intention detection and open intention discovery based on text information of human-machine dialogue based on deep learning models. Related achievements have also been published in the top academic international conferences in the field of artificial intelligence in recent years, such as *ACL, AAAI, ACM MM*, and well-known international journals, such as *Pattern Recognition* and *Knowledge-based Systems*. In order to systematically present the latest achievements in intention classification, unknown intention detection, and open intention discovery in academia in recent years, the relevant work achievements are systematically sorted out and presented to readers in the form of a complete systematic discussion.

Currently, the research on intention understanding in natural interaction develops quickly. The author's research team will timely sort out and summarize the latest achievements and share them with readers in the form of a series of books in the future. This book can not only be used as a professional textbook in the fields of natural interaction, intelligent question answering (customer service), natural language processing, human-computer interaction, etc., but also as an important reference book for the research and development of systems and products in intelligent robots, natural language processing, human-computer interaction, etc.

As the natural interaction is a new and rapidly developing research field, limited by the author's knowledge and cognitive scope, mistakes and shortcomings in the book are inevitable. We sincerely hope that you can give us valuable comments and suggestions for our book. Please contact xuhua@tsinghua.edu.cn or a third party in the open-source system platform https://thuiar.github.io/ to give us a message. All of the related source codes and datasets for this book have also been shared on the following websites https://github.com/thuiar/Books.

The research work and writing of this book were supported by the National Natural Science Foundation of China (Project No. 62173195). We deeply appreciate the following student from State Key laboratory of Intelligent Technology and Systems, Department of Computer Science and Technology, Tsinghua University for her hard preparing work: Xiaofei Chen. We also deeply appreciate the following students for the related research directions of cooperative innovation work: Ting-en Lin, Hanlei Zhang, Wenmeng Yu, and Xin Wang. Without the efforts of the members of our team, the book could not be presented in a structured form in front of every reader.

Beijing, China   Hua Xu
Beijing, China   Hanlei Zhang
Beijing, China   Ting-En Lin
November 2022

Contents

List of Figures

List of Tables

About the Authors

Hua Xu is a leading expert on Multi-modal Natural Interaction for service robots, Evolutionary Learning, and Intelligent Optimization. He is currently a Tenured Associate Professor at Tsinghua University, Editor-in-Chief of *Intelligent Systems with Applications* and Associate Editor of *Expert Systems with Application*. Prof. Xu has authored the Chinese books *Data Mining: Methodology and Applications* (2014), *Data Mining: Methods and Applications-Application Cases* (2017), *Evolutionary Machine Learning* (2021), *Data Mining: Methodology and Applications (2nd Edition)* (2022), *Natural Interaction for Tri-Co Robots (1) Human-machine Dialogue Intention Understanding* (2022), and *Natural Interaction for Tri-Co Robots (2) Sentiment Analysis of Multimodal Interaction Information* (2023), and published more than 140 papers in top-tier international journals and conferences. He is Core Expert of the No.03 National Science and Technology Major Project of the Ministry of Industry and Information Technology of China, Senior Member of CCF, member of CAAI and ACM, Vice Chairman of Tsinghua Collaborative Innovation Alliance of Robotics and Industry, and recipient of numerous awards, including the Second Prize of National Award for Progress in Science and Technology, the First Prize for Technological Invention of CFLP and First Prize for Science and Technology Progress of CFLP, etc.

Hanlei Zhang obtained his B.S. degree from the Department of Computer Science and Technology at Beijing Jiaotong University in 2020. He is currently pursuing a Ph.D. in the Department of Computer Science and Technology at Tsinghua University. His research goal focuses on analyzing human intentions in real-world scenarios. Hanlei has published five first-authored peer-reviewed papers in top-tier international conferences and journals, including *AAAI, ACM MM, ACL*, and *IEEE/ACM TASLP*. His research interests encompass various areas such as intent analysis, open world classification, clustering, multi-modal content understanding, and natural language processing.

During his undergraduate studies, Hanlei was the recipient of the National Scholarship twice and was also recognized as a Beijing Excellent graduate. During his Ph.D. career, he has received the first prize of the overall excellence scholarship twice and has been nominated for the Apple Scholars in AL/ML.

Ting-En Lin stands at the forefront of Conversational AI as a Senior Researcher at Alibaba's prestigious DAMO Academy. With an unwavering commitment to advancing the field of Natural Language Processing, He has made significant strides in Multi-modal Understanding, shaping the future of human-computer interaction.

Educated at renowned institutions, He earned his Bachelor's degree in Electrical and Computer Engineering from National Chiao Tung University (NCTU) before pursuing his M.Phil. in Computer Science and Technology at Tsinghua University (THU). Under the guidance of Prof. Hua Xu, Ting-En Lin honed his expertise and embarked on a fruitful research career.

As a prolific contributor to the AI community, He has published numerous papers in leading conferences such as *ACL*, *EMNLP*, *KDD*, *SIGIR*, and *AAAI*. He is also a dedicated member of the program committee, ensuring the growth and development of the field.

Part I
Overview

Chapter 1
Dialogue System

Abstract With the rapid development of Internet technology and the increasing popularity of intelligent robots and intelligent hardware devices, traditional keyword-based search engine information retrieval is no longer sufficient. Dialogue systems that enable users to interact naturally with robots and devices through voice dialogue and other natural language methods have gained attention and have a profound impact on daily life. Dialogue systems can be non-task-oriented or task-oriented, and can save labor costs and improve work efficiency by replacing repetitive mental work. However, building a human-computer dialogue system is challenging, especially the natural language understanding module that converts user input into structured semantic representations. Discovering unknown user intents that have not appeared in the training set and have not been identified is crucial for improving service quality and providing personalized services. Existing research on new intent discovery is divided into three schools based on unknown intent detection, non-clustering, and semi-clustering, but accurately identifying and understanding new types of user intents is still challenging. Efficiently identifying intents is a key step in answering user questions, and continuous optimization of intent recognition and discovery algorithms is necessary. The discovery of new intents in dialogue is a promising research field with high commercial value.

Keywords Dialogue systems · Structured semantic representations · Unknown intent detection · Intent recognition · Discovery algorithms

1.1 Review of Dialogue System

The proliferation of Internet technology and its widespread use has led to an unprecedented explosion of information available to individuals, growing exponentially in volume. How to enable users to quickly and accurately obtain the information they need from the massive data has become an important demand in the information age. However, with the increasing popularity of intelligent robots and intelligent hardware devices, people are no longer satisfied with the traditional search engine information retrieval mode based on keywords, but hope to obtain information through voice dialogue and other more natural ways. The emergence of

© The Author(s), under exclusive license to Springer Nature Singapore Pte Ltd. 2023
H. Xu et al., *Intent Recognition for Human-Machine Interactions*, SpringerBriefs in Computer Science, https://doi.org/10.1007/978-981-99-3885-8_1

dialogue system enables people to interact naturally with various kinds of robots and intelligent devices in the form of natural language and obtain information. This new way of interaction has not only attracted the wide attention of researchers, but also has a profound impact on people's daily life.

With the rapid development of artificial intelligence and natural language processing technology, human-computer dialogue system oriented towards inclusive natural interaction has gradually played a crucial role in daily life. Dialog system allows users to interact with machines in the form of voice or natural language text, and has a wide range of application scenarios. In addition to basic information acquisition and inquiry, it can also replace a variety of repetitive mental work, thus saving a lot of labor costs and effectively improving work efficiency. Generally speaking, the dialogue system can be divided into two types: non-task-oriented and task-oriented. The non-task-oriented dialogue system mainly consists of open domain chat or question and answer, and meets the needs of users for entertainment and service consultation through dialogue interaction. Task-oriented dialogue system can help users complete specific tasks, such as flight booking, restaurant reservation, etc., through multiple rounds of interaction to confirm user needs and complete the service. For example, intelligent outbound robot can make real-time phone calls to users for questionnaire survey. Intelligent customer service can provide all-weather online service consultation and business inquiry. Intelligent Personal Assistant (IPA) can complete corresponding tasks through real-time voice interaction. Represented by intelligent voice assistant, major enterprises have realized its commercial value and launched products with synchronous function positioning one after another [1], such as Amazon Alexa, Apple Siri, Microsoft Cortana and Google Voice Assistant abroad, and Ali Xiaomi, Baidu Xiaodu and Xiaomi Xiao'ai in China. The future prospect is unlimited.

Traditional Dialogue system is composed of multiple functional modules, including four core functional modules [2], which are Natural Language Understanding (NLU), Dialogue Management (Dialogue Management). DM), Knowledge Base (KB), and Natural Language Generation (NLG). The natural language understanding module contains three sub-tasks of domain recognition, intent recognition and slot filling, which is responsible for parsing the natural language input by users into the machine-understandable content and passing it into the dialogue management module in the form of semantic framework. The dialog management module contains two sub-tasks, such as dialog state tracking and dialog strategy learning. It is responsible for maintaining the logic, integrity, and fluency of the whole dialog, and reading or modifying the information in the knowledge base according to the dialog state. Finally, the natural language generation module outputs the corresponding response based on the dialogue strategy and the system action.

Building a human-computer dialogue system is a challenging task, especially the natural language understanding module, which is responsible for converting user input statements into a structured semantic representation that can be understood by the machine, and mapping it to the user's predefined intents so that the system can give corresponding responses. However, in reality, due to the complexity and diversity of user needs, it is difficult to design a natural language understanding module to cover all types of user intents. With the vigorous development of human-computer dialogue system, users' various demands are also increasing. How to find these unknown user intents that have never appeared in the training set and have not been identified has become a crucial problem.

By accurately identifying user intents and discovering unknown needs, the quality of existing service requirements can be further improved, and at the same time, users' interests and preferences can be deeply captured and personalized services can be provided. Therefore, it has high commercial value. New intent discovery is a new research field, and there are not many related researches. Research related to the new intent discovery is mainly divided into three schools. The first school is based on unknown intent detection beyond the scope of the domain [3, 4]. Through the modeling of known intents, the model can detect unknown intent samples that do not belong to known intents. The second school is the discovery of new intents based on non-clustering [5], which groups similar sentences with specific distance measures to discover potential new intent categories. The third school is the discovery of new intents based on semi-clustering [6], which introduces a modest quantity of annotated data as prior knowledge to guide the identification of new intents through a clustering approach. However, it is still challenging to find dialogue intents because unknown intent samples cannot be used for training or reference adjustment and unknown intents cannot be accurately identified.

In order to answer and solve user questions efficiently, it is necessary to accurately identify and understand the new types of user intents. This requires a basic knowledge of natural language processing, as well as an understanding of the intent recognition and discovery algorithm and continuous optimization. It is a key step to identify the type of intents. Therefore, it is hope to discover unknown requirements while ensuring that the known requirements of users are met. In this way, unknown intents can not only be prevented from being misclassified into known intents, but also be used to explore more potential intents.

The discovery of new intents in dialogue is a new research field, which can be regarded as the interdisciplinary research of natural language processing, unknown intent detection beyond the domain, transfer learning and clustering. In addition to the task-based conversation scenarios, the new intent discovery algorithm can be extended to the chatterbox and non-cooperative conversation scenarios. Secondly, not all user statements own specific intents, so it can be considered to detect outliers and filter noise data while discovering new types of intents based on clustering algorithm. Finally, in the actual scenario, a sentence may imply multiple intents at the same time, so the multi-label clustering model can be considered for new intent discovery. It is believed that the new intent discovery algorithm can play a greater value in the era of big data.

References

1. Chen, Y.-N., Celikyilmaz, A., Hakkani-Tur, D: Deep learning for dialogue systems. Proceedings of the 27th International Conference on Computational Linguistics: Tutorial Abstracts, pp. 25–31 (2018)
2. Chen, H., Liu, X., Yin, D., et al.: A survey on dialogue systems: recent advances and new frontiers. Assoc. Comput. Mach. Spec. Interest. Group Knowl. Discov. Data Min. Explorat. Newsletter. **19(2)**, 25–35 (2017)
3. Lin, T.E., Xu, H.: Deep unknown intent detection with margin loss. Proceedings of the 57th Annual Meeting of the Association for Computational Linguistics, pp. 5491–5496 (2019)
4. Lin, T.-E., Xu, H.: A post-processing method for detecting unknown intent of dialogue system via pre-trained deep neural network classifier. Knowl. Based Syst. **186(15)**, 104979 (2019)
5. Hakkani-Tür, D., Ju, Y.C., Zweig, G., et al: Clustering novel intents in a conversational interaction system with semantic parsing. Proceedings of the 16th Annual Conference of the International Speech Communication Association, pp. 1854–1858 (2015)
6. Lin, T.-E., Xu, H., Zhang H.: Discovering new intents via constrained deep adaptive clustering with cluster refinement. Proceedings of the 24th Association for the Advancement of Artificial Intelligence Conference on Artificial Intelligence, pp. 8360–8367 (2020)

Chapter 2
Intent Recognition

Abstract In practical scenarios of the human-machine dialogue system, intent detection is an important and challenging problem. The human-machine dialogue system firstly converts the user's query into a structured semantic representation that the machine can understand and maps it to the system's pre-defined user intents so that the system can make a corresponding reply. The process of mapping user semantics to pre-defined user intents is called "known intent classification." Accurate and efficient intent classification can guarantee that the dialogue system provides accurate services to users. With the vigorous development of human-machine dialogue systems, various new needs of users are also constantly increasing. How to discover new types of user intentions that have not been previously observed in the training set and have not yet been satisfied has become a vital question. The above process of detecting new user intents that do not appear in the pre-defined intent set is called "unknown intent detection." After successfully separating unknown intents from known intents, more attention is given to what new intents have been discovered. The process of classifying unknown intents into new user intents is called "new intent discovery." By discovering new intents of users, it can not only identify potential business opportunities but also provide guidance for the future research and development direction of the system and better understand potential user needs.

To realize an effective dialogue intent recognition process, obtaining a good intent feature representation is firstly necessary. The quality of the intent representation plays a crucial role in the performance of subsequent intent classification. Therefore, Section 2.1 briefly reviews the research on intent representation. Sections 2.2 and 2.3 introduces the literature related to known and unknown intent detection. Section 2.4 reviews the literature related to new intent discovery. Section 2.5 concludes the chapter and leads the research works in this book through the literature review.

Keywords Intent detection · Known intent classification · Unknown intent detection · New intent discovery · Human-machine dialogue system

2.1   Review of the Literature on Intent Representation

The feature representation of data is one of the core issues in machine learning. It is necessary to convert different data into machine-readable symbols for subsequent analysis tasks. How to represent textual data is also the key to natural language processing. Good textual representation plays a vital role in many natural language processing tasks, such as sentiment analysis (SA) [1], named entity recognition (NER) [2], and text generation [3]. Similarly, the effective representation of intent features is also a critical issue that must be solved in the intent detection of dialogue systems.

In natural language processing, word vectors are a common technique for textual feature representation. For complex and abstract textual data, high-quality word vectors can model the semantic relationship between different texts to extract the text's key features much better. The intent representation is described below from two aspects: "discrete representation" and "distributed representation."

2.1.1   Discrete Representation

Previous researchers used high-dimensional, sparse, and discrete vectors for text representation. This section will introduce four common discrete representation methods: one-hot encoding, bag-of-words model, Term Frequency–Inverse Document Frequency (TF-IDF), N-gram, etc.

One Hot Representation

One-hot encoding is one of the most basic and common text feature representation methods in natural language processing problems. Words are represented by high-dimensional discrete value vectors of 0 or 1, where the vector dimension is the vocabulary size. The words in the vector have the value 1 in the corresponding position of vocabulary and 0 in the rest. For example, the word "China" is represented as [0,1,0,...,0,0], Beijing is represented as [0,0,1,...,0,0], then the index of "China" in the vocabulary is 1, and "Beijing" is 2.

This simple representation method has achieved specific results in standard machine learning tasks, but there are also shortcomings. First of all, the dimension represented by one-hot encoding is proportional to the vocabulary size, which can easily cause the problem of the "curse of dimensionality" [4], and the high-dimensional spatial sparse matrix will also seriously waste computing resources. Secondly, "China" and "Beijing" should be related, but the one-hot encoded vectors are independent and cannot reflect the correlation between words. Third, one-hot encoding cannot measure the importance of words in a sentence.

Bag of Word

The bag-of-words model [3] ignores information such as the order and grammar of words and directly represents the text. The dimension of the bag-of-words model's encoding vector is also the vocabulary size. The index values of different word positions in the vocabulary in the vector are the frequency of occurrence of the word in the text.

Although this method can count the frequency of each word in the text, it also ignores the order information of the words and cannot distinguish the importance of the words in the sentence.

Term Frequency–Inverse Document Frequency (TF-IDF)

To solve the problem that the one-hot encoding and the bag-of-words model cannot distinguish the importance of words, the TF-IDF algorithm [4] came into being. Term frequency TF is the frequency of words in the document, and the high frequency tends to be essential or common words. The document frequency DF is the proportion of texts containing a specific word in the corpus. The inverse document frequency IDF is the logarithmic value of the reciprocal of the document frequency. The larger the IDF value, the less the field of words appears, and the more important they are, relatively. TF-IDF is the product of TF and IDF.

This method is simple and easy to keep important words while filtering out common words. But there is still the problem of not being able to reflect the word position information. Because it is a statistical algorithm based on the corpus, TF-IDF has relatively high requirements for the quality of the corpus.

N-gram

To overcome the problem of losing position information in the previous three methods, the N-gram [5] representation is introduced. N-gram is based on a language model. For a text sequence $\{w_1, w_2, \ldots, w_n\}$, the language model calculates the probability that this sequence is plausible.

$$p(w_1, w_2, \cdots, w_n) = \prod_{i=1}^{n} P(w_i | w_1, \cdots, w_{i-1}) \qquad (2.1)$$

Each conditional probability in the above joint probability is a parameter of the language model, and the larger the probability, the more reasonable the sentence formed.

However, the expression of the statistical language model formula (2.1) has serious flaws. For a sentence with a vocabulary size of V and a length of T, the theoretically generated sentence combinations are V^T, and the number of parameters will reach $O(TV^T)$. Due to the huge number of parameters, the probability cannot be accurately estimated. There is still a sparsity problem due to generating many combinations that do not appear in the corpus.

The N-gram language model simplifies the formula (2.1), and the probability of a word is only related to the likelihood of $m - 1$ words that appeared before it.

$$p(w_1, w_2, \cdots, w_n) = \prod_{i=1}^{n} P(w_i | w_{i-m+1}, \cdots, w_{i-1}) \tag{2.2}$$

N-gram combines two adjacent N words into a group for all the sentences in the corpus. It encodes them according to the group sequence index to obtain a high-dimensional discrete vector. For the N-gram adjacent phrase corresponding to each index position, if it appears in the text, the value of this position in the vector is 1, otherwise 0.

2.1.2 Distributed Representation

In Sect. 2.1.1, the method of discrete feature representation is introduced. The discrete feature representation method often has the problems of high computational overhead for high-dimensional vectors and lack of relevance in the textual representation. Therefore, the following briefly introduces the method of distributed feature representation.

Distributed feature representation transforms high-dimensional and sparse vectors into low-dimensional and dense vectors, which can better reflect the similarity of text representations and avoid the disaster of dimensionality. At the same time, each dimension will contain more semantic information. In 1954, Harris [6] proposed the distributional hypothesis that the similarity of words is related to the context. In 1957, Firth [7] further clarified that "the context determines the semantics of words." Based on the above theories, related researchers proposed two types of distributed feature representation methods based on a matrix and neural networks.

Matrix-Based Distributed Representation

The matrix-based distributed representation method needs to construct a co-occurrence matrix corresponding to words and contexts through a large amount of text corpus. First, selecting the word's context in combination with the specified window size is necessary. The context can be a document, a word, or an N-gram. The co-occurrence matrix is obtained by counting the number of co-occurrences of the word and the context. However, as the vocabulary increases, the dimension of the word vector will also increase, and there is still the problem of matrix sparsity. Pennington et al. proposed the Global Vector (Glove) model in 2014 [8], which trains the co-occurrence information of words through global matrix decomposition to obtain word vectors. The Glove model can obtain low-dimensional dense vector representations while considering information of word-context. The quality of word vectors is good, but it cannot be dynamically adjusted with different contexts.

Neural Network-Based Distributed Representation

In 1986, Hinton first proposed the idea of distributed representation based on the neural network [9]. In 2003, Bengio first used the neural network to train the language model (NNLM) to obtain the word vector [10]. The three-layer neural network was used to construct the language model. The neural network structure is as follows. As shown in Fig. 2.1. Given word vector matrix E, where the dimension is $|V| \times m$, $|V|$ is the vocabulary size. m is the word vector dimension. $E(w)$ represents the vectorized form of the word w. The $E(w)$ vector is stored as a row in a matrix that contains the vectorized forms of all the words in the vocabulary. The neural network's input layer is composed of $n - 1$ word vectors, where n is the sequence length of the sentence, then mapped by the neural network to obtain the hidden layer and the output layer. The dimension of the output layer is $|V|$, and the output probability of each word is normalized by the Softmax activation function.

However, the above-mentioned neural network language model can only combine limited contextual information. In order to solve this problem, Mikolov et al. [11] proposed a recurrent neural network language model (RNNLM), and the structure is shown in Fig. 2.2.

For the t-th time step, RNNLM can be expressed as:

$$x_t = [E(w_t); S_{t-1}]\qquad(2.3)$$

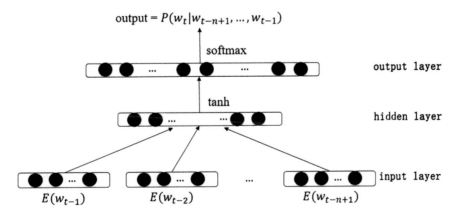

output $= P(w_t | w_{t-n+1}, \ldots, w_{t-1})$

softmax

output layer

tanh

hidden layer

input layer

$E(w_{t-1})$ $E(w_{t-2})$ $E(w_{t-n+1})$

Fig. 2.1 Neural network language model

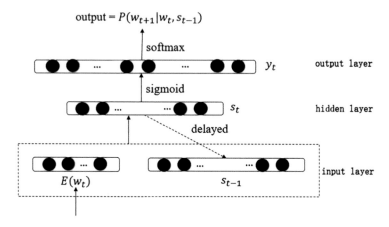

Fig. 2.2 Recurrent neural network language model

$$s_t = f(Ux_t + b) \tag{2.4}$$

$$y_t = g(VS_t + d) \tag{2.5}$$

Where U and V are the weight matrices, b and d are the biases of the hidden layer and the output layer. f represents the Sigmoid activation function and g represents the Softmax activation function. The hidden state of the current time step and the input of the next time step are used together as the initial state of the next time step. Therefore, RNNLM can effectively capture long-distance historical information and solve the problem that NNLM can only combine limited contextual information.

For both NNLM and RNNLM, the word vector is merely a by-product of training the neural network model. Due to the large number of parameters for training the neural network, the computational cost of the word vector is very high. In 2013, Mikolov et al. [12] focused on solving the word vector problem based on NNLM. By removing the hidden layer of the neural network, the training time was significantly reduced, and high-quality word vector representation was obtained. It mainly involves two essential models, namely Continuous Bag of Words Model (CBOW) and Continuous Skip-gram Model (Skip-gram); the model structure is shown in Figs. 2.3 and 2.4.

$$Loss_{CBOW} = \frac{1}{T} \sum_{t=1}^{T} \log \left(p \left(w_t \middle| \sum_{-c \le j \le c, j \ne 0} w_{t+j} \right) \right) \tag{2.6}$$

The CBOW model is divided into an input layer, projection layer, and output layer, and the goal is to predict the central word through the context. The input layer contains the context word vector $E(w_{t-c}), \ldots, E(w_{t-1}), E(w_{t+1}), \ldots, E(w_{t+c}) \in R^D$, where c is the context window size and D is the word vector dimension. The projection layer projects the context word vectors into D-dimensional vectors by

Fig. 2.3 CBOW model

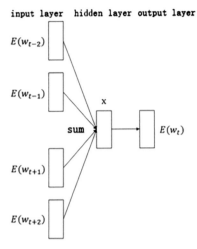

Fig. 2.4 Skip-gram model

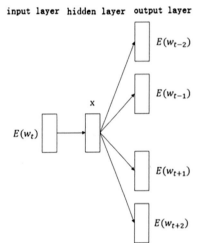

summing and averaging them. In the output layer, the traditional fully connected layer Softmax is replaced by the hierarchical Softmax using the binary tree structure, which significantly reduces the computational cost of the output layer. Since the continuous context distribution is used to represent and the order of the context word vector does not affect the results of the projection layer, it is called the continuous bag of words (CBOW). The CBOW model predicts the central word according to the context words and maximizes the log-likelihood as the objective function.

$$Loss_{skip-gram} = \frac{1}{T} \sum_{t=1}^{T} \sum_{-c \leq j \leq c, j \neq 0} \log\left(p\left(w_{t+j}|w_t\right)\right) \tag{2.7}$$

Based on the Skip-gram model, Mikolov et al. used hierarchical Softmax, negative sampling, and high-frequency word undersampling [13] to optimize the fully connected Softmax output layer to improve the convergence performance of the model further and shorten the training time. In addition, in order to verify whether the trained word vectors are semantically relevant, the CBOW model and the Skip-gram model have designed such a task. Given words with related properties such as "China" and "Beijing," "United States," and "Washington," high-quality word vectors should satisfy the property that E(China) − E(Beijing) is equal to E(United States) − E(Washington). The CBOW and Skip-gram models are better than the traditional NNLM models on this task.

Neural network-based distributed representation solves the problem of the high computational cost of discrete feature representation and the lack of correlation between vectors. However, the word vectors obtained by NNLM, RNNLM, and other methods are static. They not only cannot be dynamically adjusted in combination with different contexts, but also fail to solve polysemy problems. In order to solve this problem, researchers have proposed a series of classic pre-trained language models in recent years, which have achieved remarkable results. The following three main pre-trained language models will be introduced. The model structures are shown in Fig. 2.5.

In 2018, Peters et al. proposed ELMo model [14] (Embedding from Language Models), which can learn dynamic word vector representations from a deep bidirectional language model. The model structure is shown in Fig. 2.5a. The model uses the bidirectional long-short-term memory neural network (Bi-LSTM) to encode the context-before and the context-after vector independently. Afterward, they combine the context vectors to obtain the word vector representation with context information, then maximize the log-likelihood to obtain the training objective.

$$Loss_{ELMO} = \sum_{t=1}^{T} \left(\log p\left(w_t | w_1, \cdots, w_{t-1}; \theta_x, \overrightarrow{\theta_{LSTM}}, \theta_s\right) \right. $$
$$\left. + \log p\left(w_t | w_{t+1}, \cdots, w_T; \theta_x, \overleftarrow{\theta_{LSTM}}, \theta_s\right)\right) \tag{2.8}$$

where θ_x is the word vector parameter. $\overleftarrow{\theta_{LSTM}}$, and $\overrightarrow{\theta_{LSTM}}$ is the network parameter. θ_s is the network parameter.

In 2018, Radford et al. proposed the GPT model [15, 16], using the multi-layer Transformer [17] as a feature extractor to further improve the feature extraction capability. The model structure is shown in Fig. 2.5b. Training is divided into two stages. The first stage is the unsupervised pre-training language model, using

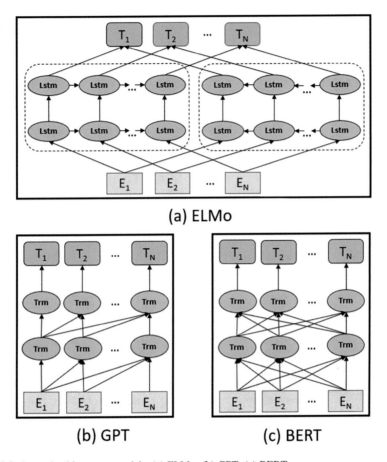

Fig. 2.5 Pre-trained language models. (**a**) ELMo. (**b**) GPT. (**c**) BERT

Transformer's decoder to train the language model, and maximizing the log-likelihood of the previous word to predict the current word to obtain the objective function.

$$Loss_{GPT} = \sum_{t=1}^{T} \left(\log p(w_t | w_{t-c}, \cdots, w_{t-1}; \theta) \right) \tag{2.2-9}$$

where c is the window size, and θ is the model parameter. The second stage uses the extracted features in the first stage to fine-tune the downstream tasks and train by maximizing the log-likelihood.

$$Loss_{finetune} = \sum_{x,y} \log p(y|x_1, x_2, \cdots, x_n) \qquad (2.10)$$

where n is the sequence length.

The GPT model uses Transformer to enhance the feature extraction capability of the model, but the obtained feature representation lacks contextual information. Therefore, in 2019, Devin et al. proposed the Bidirectional Encoder Representation from Transformers (BERT) model [18]. While using the multi-layer Transformer as an encoder, this model can consider the context information of words and has achieved superior results in several NLP tasks. The model structure is shown in Fig. 2.5c.

BERT also uses a two-stage training method of pre-training and fine-tuning. In the pre-training phase, BERT proposes two tasks, Masked Language Model (Masked LM) and Next Sentence Prediction (NSP). Masked LM aims to construct a language model and use contextual information to predict the central word. For this task, ELMo adopts the method of splicing using the context vector encoded by Bi-LSTM. However, this method cannot learn the deep information of feature representation, and directly using the context encoding vector will lead to the leakage of predicted word information. In order to enable the language model to take into account the context information while avoiding information leakage, BERT randomly replaces up to 15% of the words during the training with [mask] token. It uses an unsupervised method to predict [mask] words. However, since the masked words may never appear in the fine-tuning stage, 80% of the masked words are replaced with [mask] token, 10% are randomly replaced with other words, and the remaining 10% remain unchanged. Finally, each word vector can be fully learned. The goal of NSP is to understand the connection between two sentences. The training objective is relatively simple: for the input sentences A and B, judge whether B is the following sentence of A.

In the fine-tuning stage, BERT can be widely applied to typical NLP downstream tasks such as text classification, sequence labeling, and question-answer matching. The dialogue intent discovery task introduced in this book uses the BERT pre-trained model to capture features of intent, through downstream task fine-tuning, learns a deep intent feature representation suitable for the dialogue intent discovery task.

2.1.3   Summary

For the feature representation of intent, based on the development history of text representation, high-dimensional sparse discrete feature representation, and low-dimensional dense distributed feature representation is introduced, and the advantages and disadvantages of different feature representation methods are analyzed. The current advanced intent feature representation method is chosen as the basis to realize intent recognition.

2.2   Review of Known Intent Classification

In dialogue systems, in addition to regular intent classification, some researchers also jointly model known intent classification and slot filling, exploiting the strong correlation between the two tasks to achieve mutual benefits between the two tasks.

2.2.1   Review of Single-Model Intent Classification

Many researchers have recently proposed various models for classifying user dialogue content. For example, Kato et al. [19] proposed a data-driven autoencoder method. They reduce the total error of the autoencoder to reconstruct child nodes by building a regression tree. Kim et al. [20] proposed a method of enriching word embeddings to complete the intent classification task by forcing semantically similar or dissimilar words to be closer or farther away in the embedding space to improve the performance of the classification task. Moreover, they use different semantic dictionaries to enrich word embeddings and then use them as the initial representation of words for intent classification. Finally, they build a bidirectional LSTM based on embeddings for intent classification. Recently, with the rise of pre-trained models, the pre-training task of deep bidirectional representations using large-scale unlabeled corpora has brought improvements to various natural language processing tasks after simple fine-tuning. Among them, Chen et al. [21] explored using pre-trained BERT models for intent classification tasks, resulting in improved performance.

2.2.2   Review of Bi-model Intent Classification

Unlike the single model above, which only completes the task of intent classification, the dual model achieves mutual benefit by completing two tasks of intent classification and slot filling. In the dual model, the two tasks do not share a recurrent neural network (RNN) or convolutional neural network (CNN) but use their respective neural network structures, which are realized by sharing the hidden states of the neural networks of the two tasks. The information sharing between the two tasks can realize the purpose of joint learning. Wang et al. [22] first proposed a dual-model structure in intent classification and slot filling. Both intent classification and slot-filling tasks use their own independent encoder-decoder structure. The encoder uses a bidirectional LSTM layer, and the decoder uses a single LSTM layer. The output of the intent classification decoder at the last timestep passes through the Softmax classification layer to predict the intent. The output of the slot-filling decoder at each timestep passes through the Softmax classification layer to predict the slot label at each moment. The output of the two task encoders will be shared with the decoder of the other task at each moment, so each encoder can get not only the output of its own task encoder, but also the output of the other task encoder through this shared output. The joint learning of two tasks is realized in the way of the hidden state.

2.2.3   Summary

This chapter presents an overview of related research on known intent classification and analyzes the advantages and disadvantages of each method in details. Known intent classification research is mainly divided into two categories: the single-model-based dialog intent classification method and the bi-model-based one. This book will introduce the latest improved intent classification methods that train multiple neural networks asynchronously by sharing their internal state information.

2.3   Review of Unknown Intent Detection

By discovering unknown requirements, the quality of existing service requirements can be further improved, and at the same time, users' interests and preferences can be deeply captured, and personalized services can be provided. Therefore, it manifests high commercial values. However, due to the problems such as the inability to use unknown intent samples for training or reference, and the inability to accurately estimate the type of unknown intents, the detection of unknown intents is still challenging, and there are not many researches in related fields. This book divides unknown intent detection methods into four categories: unknown intent detection based on traditional discriminant model, unknown intent detection based on open set recognition in computer vision domain, unknown intent detection based on out-of-domain detection and unknown intent detection based on other methods.

2.3.1   Unknown Intent Detection Based on Traditional Discriminant Model

If the conversational unknown intent detection is regarded as an open classification problem and the unknown intent samples contained in the test set are regarded as one class, it is hoped that the n-class known intent samples in the training set can be used as the prior knowledge to complete the classification of n-class known types in the testing stage while effectively detecting the unknown type samples. Early researchers used one-class classifiers to solve open classification problems. One-class support vector Machine (One-class SVM) [23] uses the origin defined by kernel function as the only negative class and other classes as positive classes to maximize the spatial boundary of other classes relative to the origin. The method of support vector data description (SVDD) [24] seeks the optimal hypersphere containing most of the training samples in the feature space, and determines whether the sample is an open type depending on whether it falls into the optimal hypersphere in the feature space or not. However, this kind of method manifests poor training effect when the training data lacks negative samples [25]. One of the most widely used approaches for multi-class open classification is the classical 1-vs-rest support vector machine (1-vs-rest SVM) [26] method. For known classes $C = \{C_1, C_2, \ldots, C_n\}$ trains binary classifiers for C_i of each category. Those belonging to C_i are regarded as positive classes, and those not belonging to C_i are regarded as negative classes. The decision boundary of each class is learned through the data obtained by binary classification.

2.3.2   Unknown Intent Detection Based on Open Set Recognition in Computer Vision

Open classification is also widely used in the field of computer vision (open set identification problem). In 2013, Scheirer et al. proposed the notion of "open space risk" [27] as a benchmark for assessing open classification. Open space refers to the area where open classes are located outside the decision boundary. They extended the one-to-rest SVM method by using two parallel hyperplanes as positive boundaries to reduce open space risk, but the extent of this reduction is still relatively limited. In 2014, Jain et al. [28] obtained the classification fraction of the positive class by using the multi-class one-vs-rest RBF SVM method, and then estimated the classification probability of Weibull distribution fitting by using the extreme value theory. Scheirer et al. proposed a contraction decay probability model [29], specifically explaining how to obtain the output probability threshold of RBF one-class SVM method. However, both of the above two methods [28, 29] require unknown class samples as the decision threshold of prior knowledge selection. In 2016, Fei and Liu proposed the Support vector machine algorithm (cbsSVM) based on the center similarity [25]. This method is based on SVM and obtains the decision boundary by calculating the similarity between each class sample and the corresponding cluster center. This method also requires unknown class samples in the training stage. The above methods lack modeling deep intent representations, and the performance of open classification is limited.

Deep neural networks (DNNs) can capture the deep feature information, so researchers try to use DNNs to complete the open classification task. In 2016, Bendale and Boult proposed the OpenMax algorithm [30], which defined the specific form of open space for intent representations extracted by deep neural networks. This method uses the penultimate layer activation vector of deep convolutional neural network to sense the extreme distance, and then obtains the unknown class probability through Weibull distribution fitting. This method reduces the open space risk to a certain extent, but it still needs the unknown class sample for tuning parameters. In 2017, Shu et al. proposed the DOC algorithm [31], which uses Sigmoid activation function as the last layer of deep neural network to calculate the probability threshold based on the statistical distribution results of classified probability. Although this method proposed a method to obtain the decision boundary based on the statistical probability distribution, it failed to find the essential features to distinguish different intents.

2.3.3 Unknown Intent Detection Based on Out-of-Domain Detection

Unknown intent detection is also related to the method of out of domain (OOD) detection. In 2017, Hendrycks and Gimpel proposed a simple baseline [32] for detecting out-of-distribution samples. Related works indicated that trained samples tended to obtain higher SoftMax probability scores than untrained out-of-distribution samples. Therefore, OOD samples with low confidence can be detected by the probability threshold. In 2018, Liang et al. improved on this basis and proposed the ODIN method [33]. ODIN adopted two strategies of temperature scaling and input preprocessing to make the difference between the SoftMax probability distribution of in-domain (ID) and OOD samples larger, which made the samples with low confidence easier to be detected. However, the probability thresholds of the above two methods [32, 33] are static, and different probability thresholds have a great impact on the results, so it is difficult to combine different probability distributions for dynamic adaptive adjustment. In addition, Kim et al. [34] also tried to jointly optimize ID and OOD intent detectors. Lee et al. [35] used Mahalanobis distance and input preprocessing to train samples in the field. All the above methods [32–35] require OOD samples for parameter adjustment. In 2019, Lin and Xu [36] employed deep metric learning to extract deep intent features. They then utilized the LOF [12] outlier detection algorithm to distinguish unknown intents. However, the recognition performance of known class intents in data sets with more intent categories would be reduced.

2.3.4   Unknown Intent Detection Based on Other Methods

Brychcin and Kr′al et al. [12] tried to use clustering methods to model intent representations, but the clustering performance was not ideal and the known intents could not be effectively used as prior knowledge. In 2017, Yu et al. [37] employed reinforcement learning to produce positive and negative samples for training a classifier to differentiate between known and unknown categories. In 2018, Ryu et al. proposed an adversarial learning method based on generators and discriminators [38] to train on known class intents and judge unknown intents through discriminators. However, Nalisnick et al. [39] found that the deep generation model could not capture advanced semantic features of complex text data in the real world. The performance of discrete text data in high dimensions is often poor.

2.3.5   Summary

At present, the unknown intent detection method in dialogue scenarios through deep neural networks owns the following deficiencies: most of the existing methods need to modify the model architecture to detect the unknown intent, which cannot be directly applied to the existing intent classifiers. In addition, unknown intent samples need to be used for training or parameter adjustment, and the decision boundary is fuzzy, leading to excessive risk in open space. The method related to probability threshold is static, and different probability threshold shows great influence on the result, so it cannot be adjusted dynamically with different probability distribution. This book will introduce the latest detailed description of its improved algorithms, including the unknown intent detection method based on model post-processing, the unknown intent detection model based on large margin cosine loss function, the unknown intent detection based on deep metric learning and the unknown intent detection based on dynamic constraint boundary.

2.4   Review of New Intent Discovery

After separating intents of unknown types from those of known types, more attention is paid to what new intents are discovered. By grouping the input sentences of users through a clustering algorithm, researchers aim to automatically discover a sensible classification system for intents, including identifying new intents that were not present in the training dataset. In this section, a detailed overview of novel approaches for discovering intents using unsupervised clustering and semi-supervised clustering techniques will be presented.

2.4.1   New Intent Discovery Based on Unsupervised Clustering

In recent years, researchers have tried to discover new intents through unsupervised clustering algorithms. The typical method is to convert user statements into intent representations, and then discover new intents through K-means [40] or hierarchical clustering algorithm [41]. However, due to the fixed feature space and distance metric, traditional clustering methods are difficult to obtain ideal clustering results in high-dimensional data.

In addition to traditional clustering methods, Xu et al. [42] combined K-means algorithm with deep neural networks to propose STCC algorithm. Low dimensional and deep intent representation vectors are obtained by self-supervised neural network, which greatly improves the performance of intent representation and clustering. However, STCC algorithm can not optimize the intent representation and clustering simultaneously, so there is still space for improvement in performance. With the rapid development of deep learning, researchers have begun to study how to optimize cluster center allocation and intent feature representation simultaneously through neural networks. In 2016, Xie et al. [43] proposed the classic end-to-end deep clustering algorithm DEC. The DEC algorithm uses stack autoencoders to compress documents into dense intent representation vectors and optimizes cluster center allocation and intent representation jointly through KL divergence loss. In 2017, Yang et al. [44] improved the algorithm based on DEC and added the reconstruction error as a penalty term in the process of optimizing cluster center allocation, so as to obtain better clustering results. Chang et al. [45] proposed deep adaptive clustering algorithm DAC. The DAC algorithm transforms the clustering problem into a pairwise similarity dichotomy problem. The convolutional neural network and dynamic similarity threshold are used to judge whether the sentence pairs are similar or not, and deep clustering-friendly intent representations are learned while the cluster center allocation is optimized.

In unsupervised clustering, the main strategy is to introduce feature engineering into the clustering process to extract intent features of sentences. In 2015, Hakkani-tur et al. [46] decomposed sentences into tree structure graphs through semantic analysis, pruned and merged them according to the occurrence frequency and information entropy of subgraphs to obtain clustering results. In 2017, Brychcin et al. [47] used Gaussian mixture model to cluster intents. In 2018, Padmasundari et al. [48] improved the robustness of clustering algorithm by combining different intent representations with the results of clustering algorithm through ensemble learning strategies. Shi et al. [49] proposed AutoDial algorithm, which extracts named entities, keywords and other features in the text, and then uses hierarchical clustering algorithm to discover new intents. However, the above methods all use static intent representation vectors and cannot effectively model the context information in sentences. To sum up, the current new intent discovery methods of unsupervised clustering still have the following deficiencies. First of all, the existing model cannot effectively model the context of the intent when obtaining the intent representation. Secondly, in the absence of prior knowledge, it is difficult to obtain satisfactory intent clustering results.

2.4.2   New Intent Discovery Based on Semi-Supervised Clustering

The intent taxonomy is artificially and subjectively defined, and it is difficult to obtain satisfactory clustering results in the absence of prior knowledge. By adding prior knowledge of constraints to the clustering process, the semi-supervised clustering method can effectively improve the clustering performance and discover new intents. In this section, a comprehensive overview of the new intent discovery approach for semi-supervised clustering is presented.

First of all, the problem of discovering new intents is redefined as follows. Suppose that all the input statements of the user can correspond to a specific intents, given intents set $y_1, y_2, \ldots, y_n, y_{n+1}, \ldots, y_{n+m}$ contains n known intents and m unknown intents. How to excavate the remaining m unknown intents according to n known intents? As illustrated in Fig. 2.6, dialogue systems typically have limited labeled data and access to a vast amount of unlabeled data, which includes both known and unknown intents. However, estimating the actual number of intents in the unlabeled data can be challenging. The key is how to effectively use a small amount of annotation data to improve the clustering performance, while maintaining good generalization performance, so as to discover new intents.

Semi-supervised clustering involves incorporating a small amount of annotated data into the clustering process to improve the quality of clustering results. One common approach is to modify the objective function of the clustering algorithm to include pairwise constraints. For instance, Wagstaff et al. proposed COP-KMeans [50] by introducing Must-Link and Cannot-Link hard constraints between samples into the clustering objective function, based on the K-Means algorithm. But in a real-world scenario, not all constraints are correct, and there may be errors. Therefore,

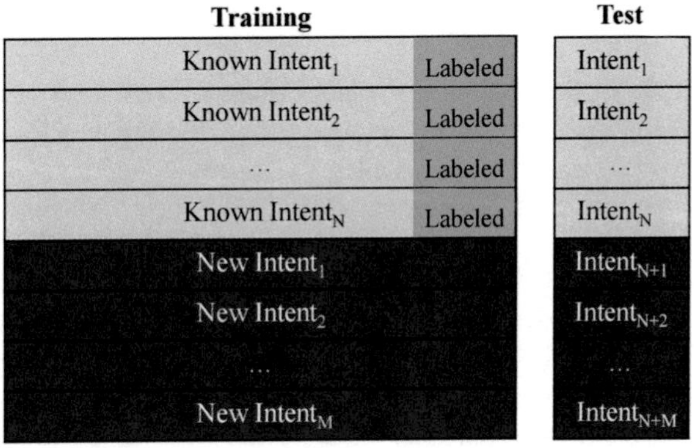

Fig. 2.6 An example for discovering new intents

Training		Test
Known Intent$_1$	Labeled	Intent$_1$
Known Intent$_2$	Labeled	Intent$_2$
...	Labeled	...
Known Intent$_N$	Labeled	Intent$_N$
New Intent$_1$		Intent$_{N+1}$
New Intent$_2$		Intent$_{N+2}$
...		...
New Intent$_M$		Intent$_{N+M}$

Basu et al. proposed the PCK-Means algorithm [51], which added soft constraints into the objective function of clustering and allowed the constraints to be broken by introducing a penalty term to make the clustering results more robust.

In addition to adding pair constraints to the clustering objective function, another method is to introduce prior knowledge through the distance metric function between samples. Bilenko et al. introduced metric learning on the basis of PCK-Means and proposed the MPCK-Means algorithm [52], so that the objective function should not only satisfy soft constraints during optimization, but also learn the distance metric function at the same time. In 2016, Wang et al. [53] extended the idea of MPCK-Means to neural networks. In addition to adding category sample constraints to the objective function, they also carried out metric learning through convolutional neural network to optimize the distance metric function.

Researchers have investigated other approaches to incorporate external supervision signals as prior knowledge that can assist the clustering process, in addition to relying on pairwise constraints and distance metric functions. In 2015, Forman et al. [54] proposed a semi-supervised new intent discovery algorithm based on user interaction, which incorporated user feedback as supervision signal in the clustering process to better discover new intents in documents. Hsu et al. [55] introduced the external similarity neural network model as prior knowledge, transferred the deep features learned from the model, and then discovered new categories in the image.

Haponchyk et al. [56] proposed the earliest method for discovering new intents of dialogues based on semi-supervised clustering in 2018, but relevant studies are still scarce at present. This method uses the pre-marked structured output template as the prior knowledge to guide the clustering process, and then finds the new intent in the dialogue through the graph cutting algorithm. In summary, the new intent discovery method by semi-supervised clustering has the following shortcomings. First of all, the prior knowledge obtained from external supervision signals has poor generalization, which easily leads to the overfitting of the model. Secondly, the existing algorithm can not jointly optimize the cluster center allocation and intent representation in the process of clustering.

2.4.3   Summary

This section presents a comprehensive evaluation of various approaches to new intent discovery, analyzing their respective advantages and disadvantages. Research on the discovery of new intents is mainly divided into two categories. The first is the new intent discovery method based on unsupervised clustering, which can directly obtain specific new intent categories. Without prior knowledge to guide the clustering process, achieving desirable clustering results can be challenging. The second type is a new intent discovery method based on semi-supervised clustering. However, the external supervision signal adopted by the existing algorithm requires a large amount of feature engineering on the data, which is not only time-consuming and laborious, but also leads to the overfitting and generalization ability of the model. In view of the shortcomings of the above intent representation and new intent discovery, the latest improved algorithm is discussed in details in the subsequent chapters of this book, including the new intent discovery model based on self-supervised constraint clustering.

2.5 Conclusion

This chapter summarizes the research on intent representation, known intent classification, unknown intent detection, new intent discovery, and analyzes the advantages and disadvantages of each method in details. In view of the shortcomings of the above methods of intent representation and dialogue intent recognition, this book will explore how to effectively use deep intent representation to carry out the task of dialogue intent discovery. In the following chapters, it will introduce two effective classification methods of known intent studied by the academic field, four unknown intent detection methods studied by us and a new intent discovery method. Book involving code at the core of the related research and methods have been released on the international open source software community platform (see the link: https://thuiar.github.io/).

The main content of this book: On the whole, intent recognition in natural interaction of intelligent robots is systematically discussed from three levels: conversational intent classification, unknown intent detection and new intent discovery. The second part of this book starts with the discussion of the most basic problem solving methods of interactive intent recognition, respectively focusing on the introduction of the intent classification model based on the single model and the intent classification model based on the double model, and in-depth comparison of the advantages and disadvantages of different intent classification methods. In the third part of this book, on the basis of the deep intent classification, focuses on the unknown intent detection in the field of text dialogue system, and further optimizes the stability of the detection algorithm. In the fourth part, based on the unknown intent detection, the author further discusses how to discover the new types of conversation intents by self-supervised clustering method according to the characteristics of unknown intent data set and learning feature representation. Finally, on the basis of the above research work, the fifth article presents an experimental demonstration platform for intent recognition of text dialogue data provided by the author in an open and shared way, which provides an important platform support for relevant personnel to carry out this work.

Summary of this part: At present, the dialogue intent recognition method, which combines the knowledge of deep learning, natural language understanding, human-computer interaction, algorithm and other fields, as the core and key technology of intelligent robot natural interaction and intelligent dialogue system, has attracted the attention of academia and industry. In the era of big data, the comprehensive use of the current data-driven deep learning method to realize the recognition of human-computer interaction intents shows potential application value for the design and implementation of autonomous intelligent interaction systems (intelligent customer service), intelligent robots, and the analysis of the potential demand and pain points of Internet products. Based on the basic methods and improved algorithms related to deep learning, this book introduces and analyzes problems, explores solutions, presents research methods, and demonstrates tool platforms from five aspects, namely, overview of intelligent dialogue system, known intent classification, unknown intent detection, new intent discovery, and demonstration experiment platform. The method and tool platform of intent recognition for human-machine natural interaction are introduced systematically, which provides a systematic and complete basic reference for the application of this technology field.

References

1. Huang, M., Qian, Q., Zhu, X.: Encoding syntactic knowledge in neural networks for sentiment classification. Assoc. Comput. Mach. Trans. Informat. Syst. **35**(3), 1–27 (2017)
2. Konkol, M., Konopík, M.: Segment representations in named entity recognition. Proceedings of the 18th International Conference on Text, Speech, and Dialogue, pp. 61–70 (2015)
3. Hajdik, V., Buys, J., Goodman, M.W., et al.: Neural text generation from rich semantic representations. Proceedings of the 7th Conference of the North American Chapter of the Association for Computational Linguistics: Human Language Technologies, pp. 2259–2266 (2019)
4. Jones, K.S.: A statistical interpretation of term specificity and its application in retrieval. Document retrieval systems, pp. 132–142 (1988)
5. Brown, P.F., Desouza, P.V., Mercer, R.L., et al.: Class-based n-gram models of natural language. Comput. Linguist. **18**(4), 467–479 (1992)
6. Harris, Z.S.: Distributional structure. Word. **10**(2–3), 146–162 (1954)
7. Firth, J.R.: A synopsis of linguistic theory, 1930–55. Studies in linguistic analysis, pp. 10–32 (1957)
8. Pennington, J., Socher, R., Manning, C.: Glove: global vectors for word representation. Proceedings of the 19th conference on empirical methods in natural language processing, pp. 1532–1543 (2014)
9. Hinton, G.E., et al.: Learning distributed representations of concepts. Proceedings of the 8th annual conference of the cognitive science society, pp. 1–12 (1986)
10. Bengio, Y., Ducharme, R.E., Vincent, P., et al.: A neural probabilistic language model. J. Mach. Learn, pp. 1137–1155 (2003)
11. Mikolov, T., Karafiát, M., Burget, L., et al.: Recurrent neural network based language model. Proceedings of the 11th annual conference of the international speech communication association, pp. 1045–1048 (2010)
12. Breunig, M.M., Kriegel, H.-P., Ng, R.T., et al.: LOF: identifying density-based local outliers. Proceedings of the 29th Special Interest Group on Management of Data international conference on Management of data, pp. 93–104 (2000)
13. Mikolov, T., Chen, K., Corrado, G., et al.: Efficient estimation of word representations in vector space. Proceedings of the 1st International Conference on Learning Representations, pp. 1–12 (2013)
14. Peters, M., Neumann, M., Iyyer, M., et al.: Deep contextualized word representations. Proceedings of the 16th Conference of the North American Chapter of the Association for Computational Linguistics: Human Language Technologies, pp. 2227–2237 (2018)
15. Radford, A., Wu, J., Child, R., et al.: Language Models are Unsupervised Multitask Learners. OpenAI Blog. (2019)

16. Radford, A., Narasimhan, K., Salimans, T., et al.: Improving language understanding by generative pre-training. OpenAI Blog. (2018)
17. Vaswani, A., Shazeer, N., Parmar, N., et al.: Attention is all you need. Proceedings of the 31st International Conference on Neural Information Processing Systems, pp. 6000–6010 (2017)
18. Devlin, J., Chang, M.-W., Lee, K., et al.: BERT: Pre-training of deep bidirectional transformers for language understanding. Proceedings of the 17th Conference of the North American Chapter of the Association for Computational Linguistics: Human Language Technologies, pp. 4171–4186 (2019)
19. Kato, T., Nagai, A., Noda, N., et al.: Utterance intent classification of a spoken dialogue system with efficiently untied recursive autoencoders. Proceedings of the 18th Annual SIGdial Meeting on Discourse and Dialogue, pp. 60–64 (2017)
20. Kim, J.K., Tur, G., Celikyilmaz, A., et al.: Intent detection using semantically enriched word embeddings. Proceedings of the Institute of Electrical and Electronics Engineers Spoken Language Technology Workshop, pp. 414–419 (2016)
21. Chen, Q., Zhuo, Z., Wang, W.: Bert for joint intent classification and slot filling. arXiv preprint arXiv:1902.10909 (2019)
22. Wang, Y., Shen, Y., Jin, H.: A bi-model based RNN semantic frame parsing model for intent detection and slot filling. Proceedings of the 16th Conference of the North American Chapter of the Association for Computational Linguistics: Human Language Technologies, pp. 309–314 (2018)
23. Schölkopf, B., Platt, J.C., Shawe-Taylor, J., et al.: Estimating the support of a high-dimensional distribution. Neural Comput. **13**(7), 1443–1471 (2001)
24. Tax, D.M.J., Duin, R.P.W.: Support vector data description. Mach. Learn. **54**(1), 45–66 (2004)
25. Fei, G., Liu, B.: Breaking the closed world assumption in text classification. Proceedings of the 16th Conference of the North American Chapter of the Association for Computational Linguistics: Human Language Technologies, pp. 506–514 (2016)
26. Rifkin, R., Klautau, A.: In defense of one-vs-all classification. J. Mach. Learn. Res. **5**, 101–141 (2004)
27. Scheirer, W.J., De Rezende Rocha, A., Sapkota, A., et al.: Toward open set recognition. Instit. Electr. Electron. Eng. Trans. Pattern Anal. Mach. Intell. **35**(7), 1757–1772 (2012)
28. Jain, L.P., Scheirer, W.J., Boult, T.E.: Multi-class open set recognition using probability of inclusion. Proceedings of the 13th European Conference on Computer Vision, pp. 393–409 (2014)
29. Scheirer, W.J., Jain, L.P., Boult, T.E.: Probability models for open set recognition. Instit. Electr. Electron. Eng. Trans. Pattern Anal. Mach. Intell. **36**(11), 2317–2324 (2014)
30. Bendale, A., Boult, T.E.: Towards open set deep networks. Proceedings of the 31st Institute of Electrical an Electronics Engineers Conference on Computer Vision and Pattern Recognition, pp. 1563–1572 (2016)
31. Shu, L., Xu, H., Liu, B.: DOC: deep open classification of text documents. Proceedings of the 22nd Conference on Empirical Methods in Natural Language Processing, pp. 2911–2916 (2017)
32. Hendrycks, D., Gimpel, K.: A baseline for detecting misclassified and out-of-distribution examples in neural networks. Proceedings of the 5th International Conference on Learning Representations (2017)
33. Liang, S., Li, Y., Srikant, R.: Enhancing the reliability of out-of-distribution image detection in neural networks. Proceeding of the 6th International Conference on Learning Representations (2018)
34. Kim, J.K., Kim, Y.B.: Joint learning of domain classification and out-of-domain detection with dynamic class weighting for satisficing false acceptance rates. Proceedings of the 19th International Speech Communication Association, pp. 556–560 (2018)
35. Lee, K., Lee, H., et al.: A simple unified framework for detecting out-of-distribution samples and adversarial attacks. Adv. Neural Inf. Proces. Syst. **31**, 7167–7177 (2018)

36. Lin, T.E., Xu, H.: Deep unknown intent detection with margin loss. Proceedings of the 57th Annual Meeting of the Association for Computational Linguistics, pp. 5491–5496 (2019)
37. Yu, Y., Qu, W.-Y., Li, N., et al.: Open-category classification by adversarial sample generation. Proceedings of the 26th International Joint Conference on Artificial Intelligence, pp. 3357–3363 (2017)
38. Ryu, S., Koo, S., Yu, H., et al.: Out-of-domain detection based on generative adversarial network. Proceedings of the 23rd Conference on Empirical Methods in Natural Language Processing, pp. 714–718 (2018)
39. Nalisnick, E., Matsukawa, A., Whye Teh, Y., et al.: Do deep generative models know what they don't know?. Proceedings of the 7th International Conference on Learning Representations (2019)
40. MacQueen, J., et al.: Some methods for classification and analysis of multivariate observations. Proceedings of the 5th Berkeley symposium on mathematical statistics and probability. pp. 281–297 (1967)
41. Gowda, K.C., Krishna, G.: Agglomerative clustering using the concept of mutual nearest neighbourhood. Pattern Recogn. **10**(2), 105–112 (1978)
42. Xu, J., Wang, P., Tian, G., et al.: Short text clustering via convolutional neural networks. Proceedings of the 1st Workshop on Vector Space Modeling for Natural Language Processing, pp. 62–69 (2015)
43. Xie, J., Girshick, R., Farhadi, A.: Unsupervised deep embedding for clustering analysis. Proceedings of the 33rd International conference on machine learning, pp. 478–487 (2016)
44. Yang, B., Fu, X., Sidiropoulos, N.D., et al.: Towards k-means-friendly spaces: simultaneous deep learning and clustering. Proceedings of the 34th International Conference on Machine Learning, pp. 3861–3870 (2017)
45. Chang, J., Wang, L., Meng, G., et al.: Deep adaptive image clustering. Proceedings of the 15th Institute of Electrical and Electronics Engineers International Conference on Computer Vision, pp. 5879–5887 (2017)
46. Hakkani-Tür, D., Ju, Y.C., Zweig, G., et al.: Clustering novel intents in a conversational interaction system with semantic parsing. Proceedings of the 16th Annual Conference of the International Speech Communication Association, pp. 1854–1858 (2015)
47. Tomas, B., Pavel, K.: Unsupervised dialogue act induction using gaussian mixtures. Proceedings of the 17th European Chapter of the Association for Computational Linguistics, pp. 485–490 (2017)
48. Padmasundari, S.B.: Intent discovery through unsupervised semantic text clustering. Proceedings of the 19th International Speech Communication Association, pp. 606–610 (2018)
49. Shi, C., Chen, Q., Sha, L., et al.: Auto-dialabel: labeling dialogue data with unsupervised learning. Proceedings of the 23rd Conference on Empirical Methods in Natural Language Processing, pp. 684–689 (2018)
50. Wagstaff, K., Cardie, C., Rogers, S., et al.: Constrained K-means clustering with background knowledge. Proceedings of the 18th International Conference on Machine Learning, pp. 577–584 (2001)
51. Basu, S., Banerjee, A., Mooney, R.J.: Active semi-supervision for pairwise constrained clustering. Proceedings of the 4th Society for Industrial and Applied Mathematics International Conference on Data Mining, pp. 333–344 (2004)
52. Bilenko, M., Basu, S., Mooney, R.J.: Integrating constraints and metric learning in semi-supervised clustering. Proceedings of the 21st international conference on Machine learning, pp. 11 (2004)
53. Wang, Z., Mi, H., Ittycheriah, A.: Semi-supervised clustering for short text via deep representation learning. Proceedings of the 20th Special Interest Group on Natural Language Learning Conference on Computational Natural Language Learning, pp. 31–39 (2016)
54. Forman, G., Nachlieli, H., Keshet, R.: Clustering by intent: a semi-supervised method to discover relevant clusters incrementally. Proceedings of the 25th European Conference on Machine Learning and Knowledge Discovery in Databases, pp. 20–36 (2015)

55. Hsu, Y.C., Lv, Z., Kira, Z.: Learning to cluster in order to transfer across domains and tasks. Proceedings of the 6th International Conference on Learning Representations. (2018)
56. Haponchyk, I., Uva, A., Yu, S., et al.: Supervised clustering of questions into intents for dialog system applications. Proceedings of the 23rd Conference on Empirical Methods in Natural Language Processing, pp. 2310–2321 (2018)

Part II
Intent Classification

Preface The main research work in this part is divided into two aspects: on the one hand, experiments are carried out on two datasets using deep neural networks, and the comparison of feedforward neural network, recurrent neural network and gated network models in small datasets. We also compare classification results on a larger dataset and a comparative introduction of standard word embedding feature representation and character-based N-gram feature representation classification methods. On the other hand, we investigate a bi-modal RNN model that performs semantic frame analysis for both intent recognition and slot-filling tasks. It contains two interrelated bidirectional long-short-term memory networks (BiLSTM) to complete intent detection and slot-filling tasks jointly. Furthermore, we analyze the impact of the decoder on model performance.

Chapter 3
Intent Classification Based on Single Model

Abstract Intent classification is an important preprocessing step in natural language processing tasks, used to categorize input text into specific intents, in order to assign them to corresponding subsystems or processing flows. In dialogue systems, intent classification is used to identify the intentions of users, so as to respond accordingly to their needs. In previous research, various types of neural network architectures have been proposed for intent classification tasks, including feedforward neural networks and recurrent neural networks (RNNs). This chapter provides a comprehensive introduction and comparison of two common RNN architectures, namely Long Short-Term Memory (LSTM) and Gated Recurrent Unit (GRU). LSTM and GRU are two common RNN architectures used for intent classification tasks and intent recognition in dialogue systems. They perform well in handling long text sequences and large dialogue datasets, and have their own advantages and applicable scenarios. The choice between these models depends on specific task requirements, dataset size, computational resources, and other factors.

Keywords Intent classification · Natural language processing · Dialogue systems · Long short-term memory · Gated recurrent unit

3.1 Introduction

In this chapter, we first compared feedforward neural networks, RNNs, LSTM networks, and GRU networks. By comparing their performance in different tasks, we explored their advantages and disadvantages. Secondly, we compared the encoding of words as character-level representations (using character sets n) with standard word vector models to mitigate out-of-vocabulary issues. Experimental results showed that in almost all cases, standard word vector models outperformed character-based word representations in classification tasks. Finally, we also explored the approach of linearly combining the scores of neural network models with the logarithm of likelihood ratio of N-gram language models to obtain the best classification results. Through these comparisons and experiments, we drew conclusions about the performance differences of different models and representation methods in classification tasks.

© The Author(s), under exclusive license to Springer Nature Singapore Pte Ltd. 2023   33
H. Xu et al., *Intent Recognition for Human-Machine Interactions*, SpringerBriefs in Computer Science, https://doi.org/10.1007/978-981-99-3885-8_3

There are two fundamental issues with N-gram-based classification methods. Firstly, the sparsity of the time range often requires a large amount of training data to achieve good generalization performance. To address data sparsity, external training data can be introduced for N-gram training [1]. However, these methods are ultimately limited by the domain-specific properties of N-gram distributions, and models trained on external data may not generalize well. In contrast, neural network language models [2] can leverage external data for word embedding training and model improvement. Deep neural network models have multiple layers that can automatically learn richer feature representations and capture more complex language relationships. This gives neural network language models an advantage in dealing with data sparsity issues. Furthermore, neural network models can learn features automatically through end-to-end training without the need for manual feature engineering, reducing the need for human intervention.

Compared to neural network language models (NNLMs) and standard N-gram language models, models based on recurrent neural networks (RNNs) and long short-term memory (LSTM) perform better in few-shot tasks because they can handle sequential data and have memory capabilities, allowing them to better capture long-term dependencies. However, comparing these RNN and LSTM-based models with neural network language models is still a puzzle, as they have different architectures and training methods. Previous comparisons of standard word embeddings in intent classification tasks also had uncertainties, hence there is a need for more in-depth research in large data environments. One possible research approach is a character N-gram-based classification method [3]. This approach captures finer-grained text features by splitting words into character-level N-gram features. In large data environments, this character N-gram-based classification method may have better performance as it can better handle sparse data and be more flexible in capturing lexical diversity and contextual relationships.

This chapter compares the performance of feedforward neural networks, recurrent neural networks, and gated network models on different datasets to investigate their performance in different contexts. A series of experiments are conducted to evaluate the effectiveness of these models in processing datasets of different scales, in order to determine which model performs best in different scenarios. In addition, standard word embedding feature representation methods are compared with character-based N-gram feature representation methods. The character-based N-gram feature representation method performs better. By comparing the performance of these two different feature representation methods in model classification performance, we discuss their advantages and disadvantages in practical applications. Through these comparisons and experimental evidence, a deeper understanding can be gained regarding the performance characteristics of various models under different datasets and feature representations, providing a strong basis for model selection and design in practical applications.

3.2   Comparison Systems

3.2.1   Baseline Systems

In this chapter, two baseline methods based on N-gram features are adopted for the experimental evaluation. One approach involves calculating the likelihood probability for each category using class-specific N-gram language models [1]. Class-specific N-gram language models are a method used in text classification tasks to compute the likelihood probability for each category. This method is based on N-gram language models, where N represents a fixed integer indicating combinations of N consecutive words in the text. The other baseline method uses the "Boostexter" boosting algorithm [4, 5]. Boostexter algorithm is an ensemble learning algorithm used for text classification tasks. The advantage of Boostexter algorithm is that it can gradually improve classification performance through an iterative process, and it has some robustness to imbalanced datasets and noisy data. By using these two baseline methods, it can provide a reference for the subsequent experimental results to better understand the performance characteristics of different models in handling different datasets and feature representations.

3.2.2   NNLM-Based Utterance Classifier

NNLM is a type of language model based on neural networks, used for predicting the next word or other labels, such as recipient labels, in text. Figure 3.1 illustrates the architecture of the NNLM baseline system. In this architecture, the input word embeddings are constructed by stacking the embedding vectors of two adjacent words, where w_1 and w_2 are neighboring words that share a projection layer P_w. This

Fig. 3.1 The standard neural network language model

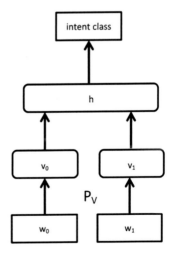

projection layer maps the embedding vectors of the two words into a single vector, which serves as the input to a multi-layer perceptron (MLP). NNLM encodes words as n-dimensional vectors, typically ranging from 100 to 1000 dimensions, depending on the task. For dialogue classification tasks, NNLM is trained to predict the class of a dialogue based on pairs of consecutive words. The overall prediction score is obtained by computing the output of the MLP. This architecture can be used not only for predicting the next word, but also for predicting other labels, such as recipient labels [6]. The introduction of NNLM as an alternative to traditional language models [7, 8] allows for better capturing of semantic relationships between words, as it uses word embedding vectors to represent words, instead of traditional one-hot encoding. NNLM has achieved good performance in natural language processing tasks and has been widely applied in text generation, sentiment analysis, machine translation, and other fields. The utterance score is calculated by the following formula.

$$
\begin{aligned}
P(L \mid w) \quad &\approx P(L_1, \ldots, L_n \mid w) = \prod_{i=1}^{n} P(L_i \mid w) \\
&\approx \prod_{i=2}^{n} P(L_i \mid w_{i-2}, w_{i-1}, h_i) = \prod_{i=1}^{n} P(L_i \mid h_i)
\end{aligned}
\tag{3.1}
$$

The best class can be calculated from $\mathrm{argmax}_L \log p(L \mid w)$.

3.2.3 RNN-Based Utterance Classifier

Recurrent Neural Network Language Model (RNNLM) [9] is a type of neural network model used for natural language modeling, which has been developed as an improvement over models based on the Markov assumption by observing that modeling the entire sentence through a series of hidden units can be more effective. Similar to Neural Network Language models [7], RNNLM also maps words to dense n-dimensional word embeddings, which can be seen as representations of words in a high-dimensional space that capture semantic relationships between words. The hidden state h_t is obtained by combining the current embedding, the previous hidden state, and the bias term through the function σ: $h_t = \sigma(W_t h_{t-1} + v_t + b_h)$. RNNLM uses a hidden state h_t to represent the internal state of the model when processing the tth word in a sentence. The hidden state h_t is calculated as a function of the current word's embedding, the previous word's hidden state h_{t-1}, and bias terms, where W_t and v_t are parameters for weight matrices and bias terms, respectively. σ denotes the activation function, commonly used ones being sigmoid, tanh, or ReLU. The hidden state h_t in RNNLM is continuously updated as each word in the sentence is processed, allowing it to capture temporal information and contextual relationships between words. This enables RNNLM to better handle long-range dependencies in language modeling tasks. Typically, the word embedding dimension in RNNLM is

Fig. 3.2 RNN classifier model

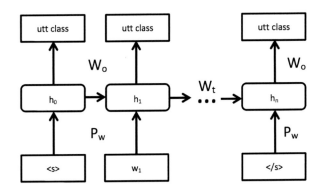

smaller, generally less than half of that in feedforward language models. This is because RNNLM preserves more contextual information through the propagation of hidden states in time, allowing for smaller values in word embedding dimensions to reduce model complexity and computational costs.

In language classification, an RNN model is used for training, as shown in Fig. 3.2. The model stores information about the current time step t in the hidden state h_t to classify the language. During testing, the probability calculation for language labels is as follows:

$$
\begin{aligned}
P(L \mid w) \quad &\approx P(L_1, \ldots, L_n \mid w) = \prod_{i=1} P(L_i \mid w) \\
&\approx \prod_{i=1}^{n} P(L_i \mid w_i, h_{i-1}) = \prod_{i=1}^{n} P(L_i \mid h_i)
\end{aligned}
\tag{3.2}
$$

This equation is reflected in the softmax output function.

3.2.4 LSTM- and GRU-Based Utterance Classifier

For a utterance classification model, the ideal scenario is to predict a unique class label for each utterance. However, in previous experiments, the performance of RNN models predicting a single label at the end of an utterance was not competitive, likely due to the issue of vanishing gradients. Therefore, the use of LSTM units for speech classification is necessary.

3.3 Experiments

3.3.1 Datasets

In this study, the ATIS corpus setting described in [8, 10] was utilized.. The training set consists of 4978 speech input samples from Class A (context-independent) section of ATIS-2 and ATIS-3, as well as 893 test speech input samples from ATIS-3 Nov93 and Dec94 datasets. The corpus contains 17 different intents, which we mapped to a binary classification task of "flight" vs. "other" (although the evaluation metrics are insensitive to prior distribution, as described below). During training, reference text transcriptions were used as labels, but during testing, ASR (Automatic Speech Recognition) outputs were used, with a word error rate of approximately 14%. There are two versions of input data: one using only the words from the original transcriptions, and another referred to as the "auto-tagged" version, where entities are replaced with phrase labels (such as city and airline) obtained from a tagger [11].

3.3.2 Experiment Settings

In previous research, different deep learning models were used with different parameter settings and word representation methods. The NNLM classifier model used 500-dimensional word embeddings and a hidden layer with 1000 hidden units, which was found to be optimal in new experiments. Initial attempts to use word hashing resulted in poor performance due to the large size of word embeddings, so word hashing was not used in this study. RNN, LSTM, and GRU models used 200-dimensional word embeddings on two corpora, and these parameter settings were shown to produce the best results for both one-hot encoding and word hashing methods. Additionally, LSTM and GRU models included a 15-dimensional hidden and memory unit layer to capture long-term dependencies. Excluding the word embeddings, the parameter count of LSTM model was about 150% higher than that of RNN, while GRU had fewer parameters than RNN. However, for large vocabulary tasks, the size of word embeddings far exceeded the number of other parameters, so only the model structure that was proven to be the most effective in experiments was used. For word hashing, the study used concurrent tri-grams and bi-grams representations, for example, the set describing the word "cat" included "#c", "ca", "at", "t#", "#ca", "cat", and "at#". This representation method may be used to construct a hash function for the vocabulary, so that hashed word representations can be used in model training and prediction.

The effects of different optimization techniques in language models that have been sufficiently trained are discussed. Due to the typical length of spoken utterances in the corpus, which is usually no more than 20 words, traditional optimization techniques such as gradient clipping, truncated backpropagation through time

(BPTT), and regularization did not significantly improve performance. Similarly, advanced optimization methods like Nesterov momentum [12] did not show noticeable performance gains. On the contrary, simple momentum methods performed well on the ATIS dataset, and using simple stochastic gradient descent yielded good results on the Cortana corpus.

This paragraph describes an experiment on intent classification using the ATIS dataset, where it was found that the final results were highly sensitive to the initial parameters. Different weight initialization strategies were used, where weights for RNN and non-gated LSTM were drawn from a normal distribution with mean 0.0 and variance 0.4, while LSTM gate weights were drawn from the same distribution but with larger gate bias values (around 5) to ensure gate values start around 1.0. To address the issue of performance variance, a heuristic approach that has shown good results in previous studies [13] was used: before training, cross-entropy was computed on a held-out set and the initialization parameters with the lowest cross-entropy from ten random seeds were selected. However, on the Cortana dataset, variance in results does not seem to be a problem, so results including error rate variance were not reported.

For the experiments on the ATIS dataset, the initial learning rate is set to 0.01 with a momentum of 3×10^{-4} for recurrent neural networks, and 0.01 with a momentum of 3×10^{-5} for LSTM models. During training, the learning rate is halved when the cross-entropy on a held-out set decreases by less than 0.01, and then continues at the same rate until the same stopping point is reached, after which the learning rate is halved at each epoch until the cross-entropy no longer decreases. Additionally, the learning rate is halved at each epoch. When selecting initial parameters, the best initial cross-entropy from 10 different initializations is chosen.

The choice of learning rate during the training process has a significant impact on the performance and convergence speed of the Cortana model, which is trained on a large dataset. Similar to the ATIS dataset, when the improvement of cross-entropy on the validation set is less than 0.1%, the learning rate is reduced. This means that if the performance improvement on the validation set is small, the learning rate will automatically decrease to avoid overly large weight updates, thereby improving the stability and performance of the model.

In model combination and evaluation, linear logistic regression (LLR) is used to calibrate scores from all models and combine multiple scores when applicable [14]. Due to the small size of the ATIS dataset, a nine-fold jack-knife cross-validation method is employed to estimate LLR parameters. Specifically, the test data is divided into nine equal-sized partitions, and the models are trained on all but one partition in turn, cycling through all partitions. Then, the scores are pooled over the entire test set, and equal error rate (EER) is used as the evaluation metric. This approach helps estimate LLR parameters and combine and evaluate models, resulting in the final performance evaluation results.

During experiments on Cortana data, the development set is used to estimate LLR weights for model combination, which are then used to combine scores from multiple models. Subsequently, the combined model is evaluated on the test set using the Equal Error Rate (EER) as the evaluation metric.

3.3.3 Experiment Results

Table 3.1 presents the results of intent classification in the ATIS dataset using different neural network architectures. The "Autotag" column refers to the condition of automatically tagging certain named entities using a lookup table. According to the results, the NNLM model performs less well compared to other models, both in standard and autotag settings. The RNN model ranks second, followed by the GRU and LSTM models, which show similar performance under different conditions. The LSTM model performs well in the standard setting, while the GRU model performs well in the autotag setting. The GRU and LSTM models outperform other methods significantly in terms of performance, and according to the results in Table 3.2, their performance difference is not significant as they fall within one standard deviation, indicating similar performance. The performance of the hash-based methods is

Table 3.1 ATIS intent classification results

System	EER(%)	Autotag EER(%)
Word 3-g LM	9.37	6.05
Word 3-g boosting	4.47	3.24
NNLM-word	6.05	4.03
RNN-word	5.26	2.45
RNN-hash	5.33	2.81
LSTM-word	2.45	1.94
LSTM-hash	2.88	2.81
GRU-word	3.24	1.58
GRU-hash	3.24	2.02

Table 3.2 Average, best held out, and oracle errors on ATIS intent classification

System	EER(%)	Autotag EER(%)
Average error		
NNLM-word	$5.83 \pm .238$	$4.15 \pm .324$
RNN-word	$4.86 \pm .919$	$3.50 \pm .775$
RNN-hash	$4.32 \pm .917$	$2.64 \pm .324$
LSTM-word	$3.38 \pm .986$	2.22 ± 1.01
LSTM-hash	4.05 ± 1.13	2.64 ± 1.02
GRU-word	4.44 ± 1.69	3.58 ± 1.24
GRU-hash	3.79 ± 1.00	2.63 ± 1.08
Oracle error		
NNLM-word	6.05(5.61)	4.03(3.60)
RNN-word	3.95(3.95)	2.45(2.45)
RNN-hash	3.59(3.24)	2.45(2.09)
LSTM-word	2.81(2.45)	2.02(1.30)
LSTM-hash	3.24(2.88)	2.02(1.22)
GRU-word	3.24(1.70)	1.30(1.30)
GRU-hash	3.24(1.30)	2.02(2.02)

Table 3.3 Cortana domain classification results

System	EER(%)	Combo EER(%)
Word 3-g LM	7.37	–
Word 3-g boosting	7.29	–
NNLM-word	9.33	7.30
RNN-word	7.99	6.80
RNN-hash	7.76	6.87
LSTM-word	6.86	6.56
LSTM-hash	7.11	6.63
GRU-word	6.78	6.46
GRU-hash	7.08	6.64

comparable to word embeddings, as they show similar performance on this small dataset.

Table 3.2 includes the mean and standard deviation results of model performance metrics. First, " Best held out " refers to the hypothetical performance of the model with the lowest cross-entropy on the validation set, selected from 10 different random seeds after training. This means that selecting the best-performing model from multiple randomly initialized models can yield more stable and reliable results. Feed-forward neural networks have much lower standard deviation compared to the lowest value among all results, but their overall performance is also the poorest. This indicates that feed-forward neural networks have smaller performance differences, i.e., lower standard deviation, but their overall performance is relatively worse. Gated networks have higher variance, while recurrent networks have lower variance, albeit with relatively smaller improvements in performance. This suggests that gated networks have larger performance differences across different random seeds, i.e., higher variance, while recurrent networks have smaller performance differences, i.e., lower variance, but their performance improvement is relatively smaller. Lastly, using the method of selecting the best seed based on the lowest cross-entropy on the validation set among 10 random seeds is a reasonable approach to selecting the best seed. Choosing the model with the lowest cross-entropy on the validation set from multiple random seeds as the best model is a reasonable choice to mitigate the impact of random initialization on experimental results.

In Table 3.3, using a larger dataset can demonstrate clearer results. The "Combo" labeled list shows the system performance when combining neural networks (NN) scores with the baseline four-gram system (FourGram). It should be noted that compared to trigram, the improvement of four-gram is limited. Similar to the ATIS dataset, the relative performance ranking of the models from worst to best is NNLM, RNN, LSTM, and GRU, with LSTM and GRU performing similarly. Word hashing performs similarly to word vectors in terms of performance, with word hashing slightly inferior for gated models and performing well for recurrent models. Character n-grams require less storage space compared to word vectors, as in this larger dataset, character n-grams only require 12 k, while words require 18 k.

Figure 3.3 shows the performance of the detection error trade-off curves in two ATIS test conditions and the Cortana task. On the large dataset (Cortana), these

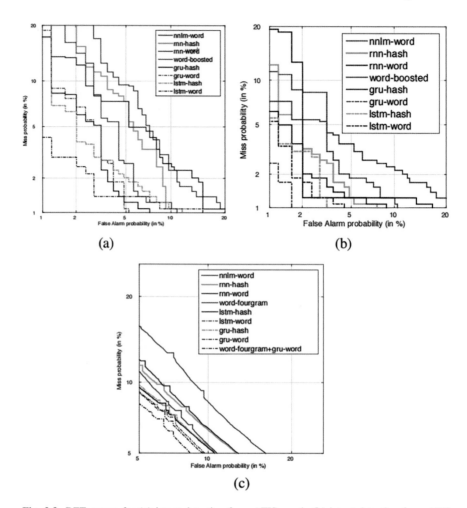

Fig. 3.3 DET curves for (**a**) intent detection from ATIS words (**b**) intent detection from ATIS autotags and (**c**) domain classification in Cortana

curves exhibit a very straight and parallel trend, indicating that there is no clear advantage or disadvantage for the various systems in different types of classification errors. This also implies that the relative performance ranking of the systems remains consistent regardless of the chosen operating point.

3.4 Conclusion

This chapter conducts research on two semantic expression classification tasks, comparing the performance of various neural network architectures on a small-scale controlled corpus (ATIS) and a large-scale real-world dataset (Cortana), and draws some conclusions. Firstly, through performance evaluation, it is found that gated recurrent networks (GRU and LSTM) perform the best in both tasks, with comparable performance and no significant differences. The next best performing architecture is conventional recurrent networks, followed by feed-forward networks. This indicates that gated recurrent networks have good performance in handling semantic expression classification tasks. Secondly, compared to standard n-gram language model (LM) classifiers or boosting classifiers, gated unit networks perform better. This implies that gated unit networks outperform traditional n-gram language models or boosting classifiers in these two tasks, possibly because gated unit networks are better at capturing long-term dependencies between input sequences. However, by combining with logistic regression, n-gram LM can further improve the neural network-based system. This suggests that combining n-gram LM with neural networks can further enhance system performance, possibly because they provide complementary information at different levels. Finally, contrary to the findings in another semantic expression classification task [13],this research did not observe consistent performance improvement from word hashing to character n-gram encoding. This means that different encoding methods may have different impacts on performance in different tasks and datasets, and the appropriate encoding method should be chosen based on the specific situation.

Current research focuses mainly on lexical information and information that can be inferred from the context of the discourse at hand. However, future research could consider incorporating non-lexical information, such as prosody, intonation, etc. [15]. This means that language models could better understand language by simulating the voice and tone variations of speakers, making text generation more natural and human-like, and closer to the way language is actually used in communication. Additionally, there is a promising model architecture feature that involves applying convolutional layers to word sequences [16]. This is a different approach from current research direction but has potential applications. Convolutional neural networks are typically used for image processing but can also be applied to text data. By applying convolutional operations on word sequences, the model can capture the correlations and local patterns between different positions, thus better understanding the contextual information in the text. This approach may show good performance in language modeling and text generation tasks and is worth further exploration for its potential applications in future research.

References

1. Lee, H., Stolcke, A., Shriberg, E.: Using out-of-domain data for lexical addressee detection in human-human-computer dialog. Proceedings of the 9th Conference of the North American Chapter of the Association for Computational Linguistics: Human Language Technologies, pp. 221–229 (2013)
2. Ravuri, S.V., Stolcke, A.: Neural network models for lexical addressee detection. Proceedings of the 15th Annual Conference of the International Speech Communication Association, pp. 741–745 (2014)
3. Huang. P.S., He, X., Gao, J., et al.: Learning deep structured semantic models for web search using clickthrough data. Proceedings of the 22nd ACM international conference on Information & Knowledge Management, pp. 2333–2338 (2013)
4. Schapire, R.E., Singer, Y.: BoosTexter: a boosting-based system for text categorization. Mach. Learn. **39**(2), 135–168 (2000)
5. Favre, B., Hakkani-Tür, D., Cuendet, S.: Icsiboost: open-source implementation of Boostexter. (2007)
6. Ravuri, S., Stolcke, A.: Neural network models for lexical addressee detection. Proceedings of the 15th Annual Conference of the International Speech Communication Association, pp. 298–302 (2014)
7. Bengio, Y., Ducharme, R., Vincent, P.: A neural probabilistic language model. J. Mach. Learn. Res. **3**, 1137–1155 (2003)
8. He, Y., Young, S.: A data-driven spoken language understanding system. Proceedings of the 9th Institute of Electrical and Electronics Engineers Workshop on Automatic Speech Recognition and Understanding, pp. 583–588 (2003)
9. Mikolov, T., Karafiát, M., Burget, L., et al.: Recurrent neural network based language model. Proceedings of the 11th annual conference of the international speech communication association, pp. 1045–1048 (2010)
10. Raymond, C., Riccardi, G.: Generative and discriminative algorithms for spoken language understanding. Proceedings of the 8th Annual Conference of the International Speech Communication Association, pp. 1605–1608 (2007)
11. Yao, K., Peng, B., Zhang, Y., Yu, D., Zweig, G., Shi, Y.: Spoken language understanding using long short-term memory neural networks, South Lake Tahoe, pp. 189–194 (2014)
12. Nesterov, Y.: A method for unconstrained convex minimization problem with the rate of convergence O $(1/k^2)$. Doklady an ussr. **269**, 543–547 (1983)
13. Ravuri, S., Stolcke, A.: Recurrent neural network and LSTM models for lexical utterance classification. Proceedings of the 16th Annual Conference of the International Speech Communication Association, pp. 135–139 (2015)
14. Pigeon, S., Druyts, P., Verlinde, P.: Applying logistic regression to the fusion of the NIST'99 1-speaker submissions. Dig. Sig. Processing. **10**(1–3), 237–248 (2000)
15. Shriberg, E., Stolcke, A., Ravuri, S.V.: Addressee detection for dialog systems using temporal and spectral dimensions of speaking style. Proceedings of the 14th Conference of the International Speech Communication Association, pp. 2559–2563 (2013)
16. Xu, P., Sarikaya, R.: Contextual domain classification in spoken language understanding systems using recurrent neural network. Proceedings of the 39th IEEE International Conference on Acoustics, Speech and Signal Processing, pp. 136–140 (2014)

Chapter 4
A Dual RNN Semantic Analysis Framework for Intent Classification and Slot

Abstract The research on spoken language understanding (SLU) system has progressed extremely fast during the past decades. Intent detection and slot filling are two main tasks for building a SLU system. Multiple deep learning based models have demonstrated good results on these tasks. The most effective algorithms are based on the structures of sequence to sequence models (or "encoder-decoder" models), and generate the intents and semantic tags either using separate models (Yao K, et al., Spoken language understanding using long short-term memory neural networks, South Lake Tahoe. 189–194, 2014; Mesnil, et al. IEEE/ACM Trans Audio Speech Lang Process, 23:530–539 2014; Peng, et al. Recurrent neural networks with external memory for spoken language understanding. Proceedings of the 2015 Natural Language Processing and Chinese Computing. 9362:25–35, 2015; Kurata, et al. Leveraging sentence-level information with encoder LSTM for semantic slot filling. Proceedings of the 2016 Conference on Empirical Methods in Natural Language Processing. 2077–2083, 2016; Hahn, et al. Inst Electr Electron Eng Trans Audio Speech Lang Process. 19:1569–1583, 2011) or a joint model (Liu and Lane. Attention-based recurrent neural network models for joint intent detection and slot filling. Proceedings of the Interspeech. 685–689, 2016; Hakkani-Tür et al., Multi-domain joint semantic frame parsing using bi-directional RNN-LSTM. Proceedings of the Interspeech. 715-719, 2016; Guo, et al. Joint semantic utterance classification and slot filling with recursive neural networks. Proceedings of the 2014 Institute of Electrical and Electronics Engineers Spoken Language Technology Workshop. 554–559, 2014). Most of the previous studies, however, either treat the intent detection and slot filling as two separate parallel tasks, or use a sequence to sequence model to generate both semantic tags and intent. Most of these approaches use one (joint) NN based model (including encoder-decoder structure) to model two tasks, hence may not fully take advantage of the cross impact between them.

In this chapter, new Bi-model based RNN semantic frame parsing network structures are designed to perform the intent detection and slot filling tasks jointly, by considering their cross-impact to each other using two correlated bidirectional LSTMs (BLSTM). The Bi-model structure with a decoder achieves state-of-the-art results on the benchmark ATIS data (Charles, et al. The atis spoken language systems pilot corpus. Proceedings of a Workshop Held at Hidden Valley. 96–101, 1990; Tur G et al. What is left to be understood in atis? IEEE Spoken Language

© The Author(s), under exclusive license to Springer Nature Singapore Pte Ltd. 2023
H. Xu et al., *Intent Recognition for Human-Machine Interactions*, SpringerBriefs in Computer Science, https://doi.org/10.1007/978-981-99-3885-8_4

Technology Workshop. 19–24, 2010), with about 0.5% intent accuracy improvement and 0.9% slot filling improvement.

Keywords Spoken language understanding · Intent detection · Slot filling · Deep learning · Joint models

4.1   Introduction

The research on SLU system has progressed extremely fast during the past decades. Two important tasks in an SLU system are intent detection and slot filling. These two tasks are normally considered as parallel tasks but may have cross-impact on each other. The intent detection is treated as an utterance classification problem, which can be modeled using conventional classifiers including regression, support vector machines (SVMs) or even deep neural. The slot filling task can be formulated as a sequence labeling problem, and the most popular approaches with good performances are using conditional random fields (CRFs) and RNN as recent works [11].

Some works also suggested using one joint RNN model for generating results of the two tasks together, by taking advantage of the sequence to sequence (or encoder-decoder) model, which also gives decent results as in literature [6].

In this chapter, Bi-model based RNN structures are proposed to take the cross-impact between two tasks into account, hence can further improve the performance of modeling an SLU system. These models can generate the intent and semantic tags concurrently for each utterance. In our Bi-model structures, two task-networks are built for the purpose of intent detection and slot filling. Each task-network includes one BLSTM with or without a LSTM decoder.

4.2   Intent Classification and Slot Filling Task Methods

4.2.1   Deep Neural Network for Intent Detection

Using deep neural networks for intent detection is similar to a standard classification problem, the only difference is that this classifier is trained under a specific domain. For example, all data in ATIS dataset is under the flight reservation domain with 18 different intent labels. There are mainly two types of models that can be used. One is a feed-forward model by taking the average of all word vectors in an utterance as its input. The other way is to use the recurrent neural network which can take each word in an utterance as a vector one by one [12].

4.2.2   Recurrent Neural Network for Slot Filling

The slot filling task is a bit different from intent detection as there are multiple outputs for the task. Hence, only RNN model is a feasible approach for this scenario. The most straight-forward way is using single RNN model generating multiple semantic tags sequentially by reading in each word one by one [2, 3, 13]. This approach has a constrain that the number of slot tags generated should be the same as that of words in an utterance. Another way to overcome this limitation is by using an encoder-decoder model containing two RNN models as an encoder for input and a decoder for output [6]. The advantage of doing this is that it gives the system capability of matching an input utterance and output slot tags with different lengths without the need of alignment. Besides using RNN, it is also possible to use the convolutional neural network (CNN) to gather with a conditional random field (CRF) to achieve slot filling task [11].

4.2.3   Joint Model for Two Tasks

It is also possible to use one joint model for intent detection and slot filling [6–8, 14, 15]. One way is to use one encoder with two decoders. The first decoder will generate sequential semantic tags and the second decoder generates the intent. Another approach is to consolidate the hidden states information from an RNN slot filling model, then generates its intent using an attention model [9]. Both of the two approaches demonstrates very good results on ATIS dataset.

4.3   Bi-Model RNN Structures for Joint Semantic Frame Parsing

Despite the success of RNN based sequence to sequence model on both tasks, most of the approaches in literature still use one single RNN model for each task or both tasks. In this section, two new Bi-model structures are proposed to take their cross-impact into account, hence further improve their performance. One structure takes the advantage of a decoder structure and the other doesn't. An asynchronous training approach based on two models' cost functions is designed to adapt to these new structures.

   A graphical illustration of two Bi-model structures with and without a decoder is shown in Fig. 4.1. The two structures are quite similar to each other except that Fig. 4.1a contains a LSTM based decoder. Hence, there is an extra decoder state s_t to be cascaded besides the encoder state h_t. The concept of using information from multiple model/multi-modal to achieve better performance has been widely used in deep learning [16–19], system identification [20–22] and also reinforcement learning

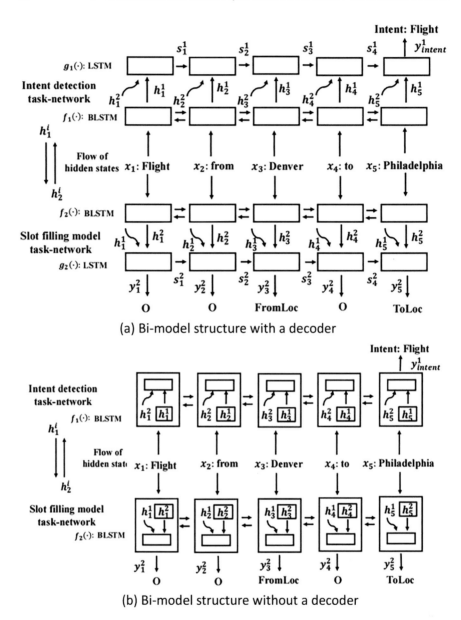

Fig. 4.1 Bi-model Structure. (**a**) Bi-model structure with a decoder. (**b**) Bi-model structure without a decoder

field recently [23, 24]. In this chapter, the work introduces a totally new approach of training multiple neural networks asynchronously by sharing their internal state information.

4.3.1 Bi-model Structure with a Decoder

The Bi-model structure with a decoder is shown as in Fig. 4.1a. There are two inter-connected bidirectional LSTMs (BLSTMs) in the structure. One is for intent detection and the other is for slot filling. Each BLSTM reads in the input utterance sequences $(x_1, x_2, \cdots, x_n)$ forward and backward, and generates two sequences of hidden states h_t^f and h_t^b. A concatenation of h_t^f and h_t^b forms a final BLSTM state $h_t = (h_t^f, h_t^b)$ at time step t. Hence, Our bidirectional LSTM $f_i(\cdot)$ generates a sequence of hidden states $(h_1^i, h_2^i, \cdots, h_n^i)$, where $i = 1$ corresponds the network for intent detection task and $i = 2$ is for the slot filling task.

In order to detect intent, hidden state h_t^1 is combined together with h_t^2 from the other bidirectional LSTM $f_2(\cdot)$ in slot filling task-network to generate the state of $g_1(\cdot)$, s_t^1, at time step t:

$$
\begin{aligned}
s_t^1 &= \phi\left(s_{t-1}^1, h_{n-1}^1, h_{n-1}^2\right) \\
y_{intent}^1 &= \arg\max_{\widehat{y}_n} P\left(\widehat{y}_n^1 \mid s_{n-1}^1, h_{n-1}^1, h_{n-1}^2\right)
\end{aligned}
\tag{4.1}
$$

where $\widehat{y}_n^1$ contains the predicted probabilities for all intent labels at the last time step n.

For the slot filling task, a similar network structure is constructed with a BLSTM $f_i(\cdot)$ and a LSTM $g_2(\cdot)$. $f_2(\cdot)$ is the same as $f_1(\cdot)$, by reading in the word sequence as its input. The difference is that there will be an output y_t^2 at each time step t for $g_2(\cdot)$, as it is a sequence labeling problem. At each step t,

$$
\begin{aligned}
s_t^2 &= \psi\left(h_{t-1}^2, h_{t-1}^1, s_{t-1}^2, y_{t-1}^2\right) \\
y_t^2 &= \arg\max_{\widehat{y}_t^2} P\left(\widehat{y}_t^2 \mid h_{t-1}^1, h_{t-1}^2, s_{t-1}^2, y_{t-1}^2\right)
\end{aligned}
\tag{4.2}
$$

where y_t^2 is the predicted semantic tags at time step t.

4.3.2 Bi-Model Structure without a Decoder

The Bi-model structure without a decoder is shown as in Fig. 4.1b. In this model, there is no LSTM decoder as in the previous model.

For the intent task, only one predicted output label y_{intent}^1 is generated from BLSTM $f_1(\cdot)$. at the last time step n, where n is the length of the utterance. Similarly, the state value h_t^1 and output intent label are generated as:

$$h_t^1 = \phi\left(h_{t-1}^1, h_{t-1}^2\right)$$
$$y_{intent}^1 = \arg\max_{\widetilde{y}_n} P\left(\widehat{y}_n^1 \mid h_{n-1}^1, h_{n-1}^2\right) \quad (4.3)$$

For the slot filling task, the basic structure of BLSTM $f_2(\cdot)$ is similar to that for the intent detection task $f_1(\cdot)$, except that there is one slot tag label y_t^2 generated at each time step t. It also takes the hidden state from two BLSTMs $f_1(\cdot)$ and $f_2(\cdot)$, i.e. h_{t-1}^1 and h_{t-1}^2, plus the output tag y_{t-1}^2 together to generate its next state value h_t^2 and also the slot tag y_t^2. To represent this as a function mathematically:

$$h_t^2 = \psi\left(h_{t-1}^2, h_{t-1}^1, y_{t-1}^2\right)$$
$$y_t^2 = \arg\max_{\widetilde{y}_t^2} P\left(\widehat{y}_t^2 \mid h_{t-1}^1, h_{t-1}^2, y_{t-1}^2\right) \quad (4.4)$$

4.3.3 Asynchronous Training

One of the major differences in the Bi-model structure is its asynchronous training, which trains two task-networks based on their own cost functions in an asynchronous manner. The loss function for intent detection task-network is $\mathcal{L}_1$, and for slot filling is cross entropy as: $\mathcal{L}_2$. $\mathcal{L}_1$ and $\mathcal{L}_2$ are defined as the following.

$$\mathcal{L}_1 \triangleq -\sum_{i=1}^{k} \widehat{y}_{intent}^{1,i} \log\left(y_{intent}^{1,i}\right) \quad (4.5)$$

and

$$\mathcal{L}_2 \triangleq -\sum_{j=1}^{n}\sum_{i=1}^{m} \widehat{y}_j^{2,i} \log\left(y_j^{2,i}\right) \quad (4.6)$$

where k is the number of intent label types, m is the number of semantic tag types and n is the number of words in a word sequence. In each training iteration, both intent detection and slot filling networks will generate a group of hidden states h_1 and h_2 from the models in previous iteration. The intent detection task-network reads in a batch of input data x_i and hidden states h_2, and generates the estimated intent labels y_{intent}^1. The intent detection task-network computes its cost based on function $\mathcal{L}_1$ and trained on that. Then the same batch of data x_i will be fed into the slot filling task network together with the hidden state h_1 from intent task-network, and further generates a batch of outputs y_i^2 for each time step. Its cost value is then computed based on cost function $\mathcal{L}_2$, and further trained on that.

The reason of using asynchronous training approach is the importance of keeping two separate cost functions for different tasks. Doing this manifests two main advantages:

It filters the negative impact between two tasks in comparison to using only one joint model, by capturing more useful information and overcoming the structural limitation of one model.

The cross-impact between two tasks can only be learned by sharing hidden states of two models, which are trained using two cost functions separately.

4.4 Experiments

4.4.1 Datasets

In this section, our proposed Bi-model structures are trained and tested on two datasets. One is the public ATIS dataset [9] containing audio recordings of flight reservations, and the other is our self-collected dataset in three different domains: Food, Home and Movie. The ATIS dataset used in this research follows the same format as in [6, 11, 13, 15]. The training set contains 4978 utterance and the test set contains 893 utterance, with a total of 18 intent types and 127 slot labels.

4.4.2 Experiment Settings

The layer sizes for both the LSTM and BLSTM networks in our model are chosen as 200. Based on the size of our dataset, the number of hidden layers is chosen as 2 and Adam optimization is used as in [25]. The size of word embedding is 300, which are initialized randomly at the beginning of experiment.

4.4.3 Experiment Results

Our first experiment is conducted on the ATIS benchmark dataset, and compared with the current existing approaches, by evaluating their intent detection accuracy and slot filling F1 scores. A detailed comparison is given in Table 4.1. Some of the models are designed for single slot filling task. Hence, only F1 scores are given. It can be observed that the new Bi-model structures outperform the current state-of-the-art results on both intent detection and slot filling tasks, and the Bi-model with a decoder also outperform that without a decoder on our ATIS dataset. The current Bi-model with a decoder shows the state-of-the-art performance on ATIS benchmark dataset with 0.9% improvement on F1 score and 0.5% improvement on intent accuracy.

Table 4.1 Performance of different models on ATIS dataset

Model	F1 score	Intent accuracy
Recursive NN [8]	93.96%	95.4%
Joint model with recurrent intent and slot label context [14]	94.47%	98.43%
Joint model with recurrent slot label context [14]	94.64%	98.21%
RNN with LABEL SAMPLING [13]	94.89%	NA
Hybrid RNN [2]	95.06%	NA
RNN-EM [3]	95.25%	NA
CNN CRF [11]	95.35%	NA
Encoder-labeler deep LSTM [4]	95.66%	NA
Joint GRU model (W) [15]	95.49%	98.10%
Attention encoder-decoder NN [6]	95.87%	98.43%
Attention BiRNN [6]	95.98%	98.21%
Bi-model without a decoder	96.65%	98.76%
Bi-model with a decoder	96.89%	98.99%

It is worth noticing that the complexities of encoder-decoder based models are normally higher than the models without using encoder decoder structures, since two networks are used and more parameters need to be updated. This is another reason why we use two models with/without using encoder-decoder structures to demonstrate the new bi-model structure design. It can also be observed that the model with a decoder gives a better result due to its higher complexity.

It is also shown in the table that the joint model in [6, 13] achieves better performance on intent detection task with slight degradation on slot filling, so a joint model is not necessary always better for both tasks. The bi-model approach overcomes this issue by generating two tasks' results separately.

Despite the absolute improvement of intent accuracy and F1 scores are only 0.5% and 0.9% on ATIS dataset, the relative improvement is not small. For intent accuracy, the number of wrongly classified utterances in test dataset reduced from 14 to 9, which gives us the 35.7% relative improvement on intent accuracy. Similarly, the relative improvement on F1 score is 22.63%.

4.5   Conclusion

In this chapter, a novel Bi-model based RNN semantic frame parsing model for intent detection and slot filling is introduced. Two substructures are discussed with the help of a decoder or not. The Bi-model structures achieve state-of-the art performance for both intent detection and slot filling on ATIS benchmark data, and also surpass the previous best SLU model on the multi-domain data. The Bi-model based RNN structure with a decoder also outperforms the Bi-model structure without a decoder on both ATIS and multi-domain data.

Summary of this Part: In recent years, research on intent classification tasks in dialogue systems has received extensive attention from academia and industry. This article systematically introduces the intent classification methods in current dialogue systems from two aspects: single-model-based and bi-model-based methods.

First, for the single-model approach, the experimental results of different deep neural networks on different datasets are explored in Chap. 3. Standard word embedding features are compared with character-based N-gram feature representations. Second, the common intent classification joint model combines dialogue intent classification and slot filling. Therefore, in this chapter, a new dual-model structure is introduced to improve further the performance of the two tasks considering the cross-effect of the two tasks and further explore the impact of the presence or absence of a decoder on performance.

In experiments with single-model intent classification, gated recurrent networks (GRU and LSTM) have the highest accuracy. N-gram language models can be combined with logistic regression to improve neural network-based models further.

In the experiments of bi-model dialogue intent classification, the complexity of the encoder-decoder-based model is generally higher than that of the model without the encoder-decoder structure due to the use of two networks - since more parameters need to be updated. This is why this chapter introduces two models with and without the encoder-decoder structure to demonstrate the design of the new dual-model structure. Furthermore, it can also be observed that the model with a decoder achieves better results due to its higher complexity.

So far, we have introduced the mainstream methods and implementation technologies for classifying known types of dialogue intent. The next chapter will introduce the detection method of unknown intent based on the intent classification introduced in this chapter.

References

1. Yao, K., Peng, B., Zhang, Y., Yu, D., Zweig, G., Shi, Y.: Spoken language understanding using long short-term memory neural networks, South Lake Tahoe, pp. 189–194 (2014)
2. Mesnil, G., Dauphin, Y., Yao, K., et al.: Using recurrent neural networks for slot filling in spoken language understanding. IEEE/ACM Trans. Audio Speech Lang. Process. **23**(3), 530–539 (2014)
3. Peng, B., Yao, K., Jing, L., et al.: Recurrent neural networks with external memory for spoken language understanding. Proceedings of the 4th Natural Language Processing and Chinese Computing, pp. 25–35 (2015)
4. Kurata, G., Xiang, B., Zhou, B., et al.: Leveraging sentence-level information with encoder LSTM for semantic slot filling. Proceedings of the 21th Conference on Empirical Methods in Natural Language Processing, pp. 2077–2083 (2016)
5. Hahn, S., Dinarelli, M., Raymond, C., et al.: Comparing stochastic approaches to spoken language understanding in multiple languages. Inst. Electr. Electron. Eng. Trans. Audio, Speech Lang. Process. **19**(6), 1569–1583 (2011)

6. Liu, B., Lane, I.: Attention-based recurrent neural network models for joint intent detection and slot filling. Proceedings of the 17th International Speech Communication Association Conference, pp. 685–689 (2016)
7. Hakkani-Tür, D., Tür, G., Celikyilmaz, A., et al.: Multi-domain joint semantic frame parsing using bi-directional RNN-LSTM. Proceedings of the 17th International Speech Communication Association Conference, pp. 715–719 (2016)
8. Guo, D., Tur, G., Yih, W., et al.: Joint semantic utterance classification and slot filling with recursive neural networks. Proceedings of the 6th Institute of Electrical and Electronics Engineers Spoken Language Technology Workshop, pp. 554–559 (2014)
9. Hemphill, C.T., Godfrey, J.J., Doddington, G.R., et al.: The atis spoken language systems pilot corpus. Proceedings of the Speech and Natural Language, pp. 24–27 (1990)
10. Tur, G., Hakkani-Tür, D., Heck, L.: What is left to be understood in atis? IEEE Spoken Language Technology Workshop, pp. 19–24 (2010)
11. Xu, P., Sarikaya, R.: Convolutional neural network based triangular crf for joint intent detection and slot filling. IEEE Workshop on Automatic Speech Recognition and Understanding, pp. 78–83 (2013)
12. Xu, P., Sarikaya, R.: Contextual domain classification in spoken language understanding systems using recurrent neural network, Proceedings of the 39th IEEE International Conference on Acoustics, Speech and Signal Processing, pp. 136–140 (2014)
13. Liu, B., Lane, I: Recurrent neural network structured output prediction for spoken language understanding. Proceedings of the 1st Neural Information Processing Systems Workshop on Machine Learning for Spoken Language Understanding and Interactions (2015)
14. Liu, B., Lane, I.: Joint online spoken language understanding and language modeling with recurrent neural networks. Proceedings of the 17th Annual Meeting of the Special Interest Group on Discourse and Dialogue, pp. 22–30 (2016)
15. Zhang, X.D., Wang, H.F.: A joint model of intent determination and slot filling for spoken language understanding. Proceedings of the 25th International Joint Conference on Artificial Intelligence, pp. 2993–2999 (2016)
16. Dean, J., Corrado, G., Monga, R., et al.: Largescale distributed deep networks. Proceedings of the 25th International Conference on Neural Information Processing Systems, pp. 1223–1231 (2012)
17. Wang, Y.: A new concept using LSTM neural networks for dynamic system identification. Proceedings of the 35th American Control Conference, pp. 5324–5329 (2017)
18. Ngiam, J., Khosla, A., Kim, M., et al.: Multimodal deep learning. Proceedings of the 28th international conference on machine learning, pp. 689–696 (2011)
19. Srivastava, N., Salakhutdinov, R.R.: Multimodal learning with deep Boltzmann machines. Proceedings of the 26th Annual Conference on Neural Information Processing Systems, pp. 2231–2239 (2012)
20. Johansen, T.A., Murray-Smith, R.: Multiple Model Approaches to Nonlinear Modelling and Control. CRC Press, Boca Raton, FL (1997)
21. Narendra, K.S., Wang, Y., Chen, W.: Stability, robustness, and performance issues in second level adaptation. Proceedings of the 32nd American Control Conference, pp. 2377–2382 (2014)
22. Narendra, K.S., Wang, Y., Chen, W.: Extension of second level adaptation using multiple models to SISO systems. Proceedings of the 33rd American Control Conference, pp. 171–176 (2015)
23. Narendra, K.S., Wang, Y., Mukhopadhay, S.: Fast reinforcement learning using multiple models. Proceedings of the 55th Institute of Electrical and Electronics Engineers Conference on Decision and Control, pp. 7183–7188 (2016)
24. Wang, Y, Jin, H.: A boosting-based deep neural networks algorithm for reinforcement learning. Proceedings of the 36th Annual American Control Conference, pp. 1065–1071 (2018)
25. Kingma, D., Ba, J.: Adam: A method for stochastic optimization. Proceedings of the 3rd International Conference on Learning Representations 2015, (2014)

Part III
Unknown Intent Detection

Preface Unknown intent detection refers to the discovery of intents that do not belong to any of the types during the classification of intents. Such intents are collectively known as unknown intents, that is unknown types of intents. The detection of unknown intents is essentially an open intent classification problem. The difficulty of unknown intent detection lies in how to realize the deep feature learning representation in open intent classification and how to learn the optimal open intent decision boundary. To solve the first difficult problem, the book explores two unknown intent detection methods in depth.

Firstly, a general post-processing method of unknown intent detection is introduced, which combines the unknown intent detection algorithm based on probability threshold and local density to make joint prediction. Secondly, the unknown intent detection method based on deep metric learning is introduced. It replaces the traditional cross entropy loss number by introducing cosine classification and large margin cosine loss function, forcing the model to consider the marginal term when learning representations, minimizing the intra-class variance while maximizing the inter-class variance. It enhances the relationship between different known classes, making the unknown intent easier to detect. The second method involves defining a spherical decision boundary to identify unknown intents and incorporating a novel loss function to automatically learn decision boundaries specific to each known class, based on the intent feature space. Samples that fall outside the learned decision boundaries are classified as unknown intents.

Chapter 5
Unknown Intent Detection Method Based on Model Post-Processing

Abstract With the maturity and popularity of dialogue systems, detecting user's unknown intent in dialogue systems has become an important task. It is also one of the most challenging tasks, as it is difficult to obtain examples, prior knowledge or the exact numbers of unknown intents. In this chapter, we introduce SofterMax and deep novelty detection (SMDN), a simple yet effective post-processing method for detecting unknown intent in dialogue systems based on pre-trained deep neural network classifiers. Our method can be flexibly applied on top of any classifiers trained in deep neural networks without changing the model architecture. By calibrating the confidence of the softmax outputs to compute the calibrated confidence score (i.e., SofterMax) and use it to calculate the decision boundary for unknown intent detection. Furthermore, the feature representations learned by the deep neural networks are fed into a traditional novelty detection algorithm to detect unknown intents from different perspectives. Finally, the methods mentioned above are combined to facilitate joint prediction. The proposed method classifies examples that differ from known intents as unknown and does not require any examples or prior knowledge of it. Extensive experiments have been conducted on three benchmark dialogue datasets. The results show that our method can yield significant improvements compared with the state-of-the-art baselines.

Keywords Novelty detection · Open-world classification · Probability calibration · Platt scaling · Dialogue system · Deep neural network

5.1 Introduction

In the first part of this book, the research progress of new intent discovery in dialogue is introduced in details. The first step in discovering new intents is to separate the known intent from the new intent, so that the classifier can correctly identify the known intents. At the same time, it can also identify new intent samples that are beyond its processing scope. However, the existing methods all need to make certain adjustments to the structure of the deep neural network model to detect new intents. New intent detection performance largely depends on whether the classifier can effectively model known intents. In contrast, traditional classification models (such as Support Vector Machines) own limited ability to model high-order semantic concepts of intents, leading to poor performance.

To solve the above problems, this chapter introduces a new intent detection method based on model post-processing, which can make the classifier model capable of new intent detection without any modification of the model structure. It can also be applied to any deep neural network classification and take full advantage of its powerful feature extraction capabilities to improve the performance of new intent detection algorithms. The method is mainly divided into two parts: the first is to calibrate the confidence of the sample output by the classifier through introducing SofterMax activation function to obtain a reasonable probability distribution. The second module combines the intent representation obtained from the deep neural network with a density-based anomaly detection algorithm for detection. Finally, the scores calculated by the above two parts are converted into probabilities through Plat Scaling for joint new intent prediction. In the absence of prior knowledge and samples about new intents, the introduced method can still detect new intents. The code is now available at https://github.com/thuiar/Books.

The rest of this chapter is organized as follows. Section 5.2 introduces the new intent detection method (SMDN) based on model post-processing and describes its sub-modules in details. It includes the introduced SofterMax activation function and deep novelty detection module. Section 5.3 describes the experimental dataset, evaluation metrics, Experimental results and analysis. Section 5.3 is the conclusion of this chapter.

5.2   A Post-Processing for New Intent Detection

This section provides a detailed description of the SofterMax and deep novelty detection methods, along with their corresponding experiment flowcharts for unknown intent detection in both single-turn and multi-turn dialogue systems (Figs. 5.1 and 5.2, respectively).To start with, we train an intent classifier based on

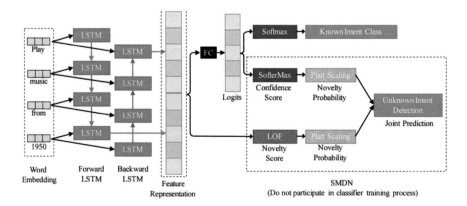

Fig. 5.1 The experiment flowchart for unknown intent detection in the single-turn dialogue system

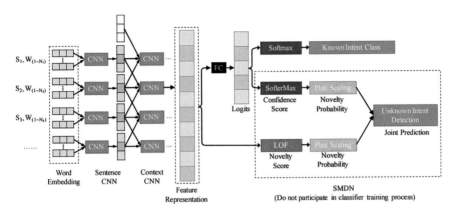

Fig. 5.2 The experiment flowchart for unknown intent detection in the multi-turn dialogue system

deep neural networks and take it as the prerequisite. Then, we calibrate the predicted confidence of the classifier through temperature scaling and then tighten the decision boundary of calibrated softmax (SofterMax) for detecting unknown intent. Also, we feed the feature representations learned by the deep neural network to the density-based novelty detection method, LOF, to discover unknown intents from different perspectives. Finally, we transform the confidence score of SofterMax and novelty score of LOF into novelty probability through Platt scaling [1] and make the joint prediction.

5.2.1 Classifiers

The key idea of the method is to detect unknown intents based on the existing intent classifier without modifying any network architecture. If an example is different from all known intents, it will be classified as the unknown. Because we detect the unknown intent based on the classifier, the performance of the intent classifier is crucial. The better the classifier we have, the better the result we get. Therefore, we implement classifiers similar to the state-of-the-art models of each dataset and compare different detection methods under the same classifier.

BiLSTM

For modeling intents of single-turn dialogue dataset, SNIPS and ATIS, we use simplified structure described in [2] and [3]. We use bidirectional LSTM (BiLSTM) [3] followed by a fully-connected layer and activation layer. BiLSTM is a many-to-one design where the dropout applies to the non-recurrent connections. We illustrate the experiment flowchart in Fig. 5.1. We use bidirectional LSTM to learn the feature representation of the intent. Then, we feed feature representation to fully-connected

(FC) layer to obtain logits. Within the dashed box is the SMDN method. We first train and validate the classifier on known samples, then test it with samples mixed with known and unknown intents.

To generate feature representations for an utterance with a maximum word sequence length of ℓ, we first convert the sequence of input words $w_{1 : \ell}$ into a sequence of m-dimensional word embeddings $x_{1 : \ell}$. These embeddings are then utilized by a forward and backward LSTM to produce the final feature representations h.

$$\overrightarrow{h_t} = LSTM\left(x_t, \overrightarrow{c_{t-1}}\right) \tag{5.1}$$

$$\overleftarrow{h_t} = LSTM\left(x_t, \overleftarrow{c_{t+1}}\right) \tag{5.2}$$

$$h = \left[\overrightarrow{h_l}; \overleftarrow{h_1}\right], z = Wh + b \tag{5.3}$$

The input at time step t is represented by $x_t \in R^m$, which is the m-dimensional word embedding. The forward and backward LSTM produce output vectors $\overrightarrow{h_t}$ and $\overleftarrow{h_t}$, respectively, while the cell state vectors are represented by $\overrightarrow{c_t}$ and $\overleftarrow{c_t}$. The output of the fully connected layer is denoted by logit z, which has the same number of neurons as the known classes. The sentence representation learned by BiLSTM is a concatenation of the last output vector of the forward LSTM $\overrightarrow{h_l}$ and the first output vector of the backward LSTM $\overleftarrow{h_1}$, represented by h. This captures the high-level semantic concepts learned by the model. We use h as the input for LOF and z as the input for Softmax.

CNN + CNN

For modeling intents of multi-turn dialogue dataset, SwDA, we use hierarchical CNN (CNN + CNN) similar to the structure described in [4]. We illustrate the experiment flowchart in Fig. 5.2. Given a sentence in the conversation, we set the maximum length of the word sequences to ℓ and the context window size of sentence sequences to c. To begin with, we perform convolution operations on word sequences to produce features as the following.

$$s_t = \text{ReLU}\left(W_{f1} \cdot x_{t:t+n-1} + b_{f1}\right) \tag{5.4}$$

We have a sentence with word embeddings $x_t \in \mathbb{R}^m$ for each of the tth words. The ReLU function is applied with a bias term b_{f1}. The filter $W_{f1} \in \mathbb{R}^{n \times m}$ performs a convolution operation on a window of n consecutive words, producing the feature s_t. We then apply the filter W_{f1} to all possible windows of words $\{x_{1:n}, x_{2:n+1}, \ldots, x_{\ell_1 - n+1:\ell_1}\}$ in the sentence, producing feature maps:

$$s = [s_1, s_2, \ldots, s_{\ell_1 - n + 1}] \tag{5.5}$$

To obtain the particular feature $\hat{s}$ associated with filter W_{f1}, a max-pooling operation is performed on the feature maps. The value of s is a real number within the set of $\mathbb{R}^{\ell_1 - n + 1}$:

$$\hat{s} = m\{s\} \tag{5.6}$$

To obtain sentence representations, we first apply a convolution operation to a matrix of word embeddings with a filter of size k_1, resulting in a feature map. We then apply a max-pooling operation to the feature map to obtain a scalar feature $\hat{s}$, which is learned by the weight matrix W_{f1}. By repeating this process with k_1 different filters, we obtain a set of scalar features that capture different aspects of the input sentence. Finally, we concatenate the scalar features to obtain the sentence representation z. This process can be repeated with multiple layers to capture increasingly complex patterns in the input.

$$\mathbf{z} = [\hat{s}_1, \ldots, \hat{s}_{k_1}] \tag{5.7}$$

Let $\mathbf{z} \in \mathbb{R}^{k_1}$ be a vector representing a sentence in k_1-dimensions.

To further generate context representations for the target sentence, we apply another convolution operation on a window of c sentences.

$$Z = [\mathbf{z}_{t-c-1}, \ldots, \mathbf{z}_{t-1}, \mathbf{z}_t, \mathbf{z}_{t+1}, \ldots, \mathbf{z}_{t+c-1}] \tag{5.8}$$

$$h_t = \text{ReLU}(W_{f2} \cdot Z_{t:t+n-1} + b_{f2}) \tag{5.9}$$

Let $Z \in \mathbb{R}^{2c-1 \times k_1}$ be a matrix that represents the sentence representations within a context window of size c for the tth sentence in a conversation. We then apply a filter $W_{f2} \in \mathbb{R}^{n \times k_1}$ to perform a convolution operation on a window of n consecutive sentences, generating the feature h_t. This filter is applied to all possible windows of sentences to produce feature maps.

$$h = [h_{t-c-1}, \ldots, h_{t-1}, h_t, h_{t+1}, \ldots, h_{t+c-1}] \tag{5.10}$$

To obtain the specific feature $\hat{h}$ for the filter W_{f2}, we perform max-pooling operation on the feature maps. Here, $h \in \mathbb{R}^{2c-1}$.

$$\hat{h} = m\{h\} \tag{5.11}$$

To obtain the context representations $\mathbf{r}$, we use k_2 different filters and apply convolution operations after learning a scalar feature $\hat{h}$ through the weight matrix W_{f2}:

$$\mathbf{r} = \left[\hat{h}_1, \ldots, \hat{h}_{k_2}\right], z = W\mathbf{r} + b \qquad (5.12)$$

To utilize the proposed method on various pre-trained deep neural network classifiers, we adopt a k_2-dimensional context vector $\mathbf{r}$ that represents the target sentence in $\mathbb{R}^{k_2}$. We also use logits z as the output of a fully connected layer, which has no softmax activation, and the number of neurons is the same as the known classes. Then, we utilize $\mathbf{r}$ as the input of the LOF layer, while z is the input of the softmax layer. This method is illustrated in both Figs. 5.1 and 5.2.

5.2.2   SofterMax

We can enhance the probability distributions obtained from the pre-trained classifier by performing confidence calibration on the softmax outputs. Following calibration, we can further tighten the decision boundary of the calibrated softmax method, which we refer to as SofterMax. This tightened boundary allows for the rejection of unknown examples. The difference between Softmax and SofterMax can be observed in Fig. 5.3.

It has been demonstrated in DOC [5] that reducing the open space risk in probability space can make it possible to reject examples for unknown classes. However, deep neural network classifiers often produce overconfident softmax output probabilities. This not only increases the open space risk by misclassifying examples belonging to unknown classes as known classes with high confidence, but also fails to provide a reasonable probability representation for output classes. Cross-entropy loss is designed to minimize the probability of other classes as much as possible, so the probability of the most likely class will always receive the highest value while the probabilities of other classes will be close to zero. Therefore, calculating per-class decision thresholds to detect unknown intent is not ideal. To address this issue, temperature scaling, proposed in [6], can be used to generate soft labels for student networks by distilling knowledge in neural networks.

Fig. 5.3 The difference between Softmax and SofterMax

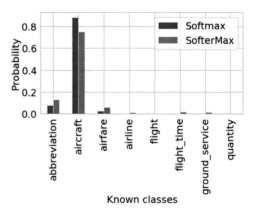

By applying temperature scaling to soften the output of Softmax, the entropy can be raised and a more reasonable probability distribution can be obtained.

Temperature Scaling

Given an input $\mathbf{x_i}$ and a neural network trained for N-class classification, the network produces a vector of logits $\mathbf{z_i}$. The class prediction $\hat{y}_i$ for input $\mathbf{x_i}$ is then obtained by computing the argmax of the vector $\mathbf{z_i}$. To obtain the confidence score $\hat{p}_i$, the softmax function σ_{SM} is applied to the logits vector $\mathbf{z_i}$. Specifically, the softmax function maps the logits vector $\mathbf{z_i}$ to a probability distribution over the N classes, and the value of $\hat{p}_i$ for a given class is the probability assigned to that class by the softmax function. Thus, the confidence score $\hat{p}_i$ represents the network's estimate of the probability that input $\mathbf{x_i}$ belongs to the predicted class $\hat{y}_i$.

$$\sigma_{SM}(\mathbf{z_i})^{(n)} = \frac{\exp\left(z_i^{(n)}\right)}{\sum_{j=1}^{N} \exp\left(z_i^{(j)}\right)} \tag{5.13}$$

$$\hat{p}_i = \max_n \sigma_{SM}(\mathbf{z_i})^{(n)} \tag{5.14}$$

We introduce the SofterMax function, denoted by $\hat{\sigma}_{SM}$, and the corresponding soften confidence scores, denoted by $\hat{q}_i$. These are defined as follows.

$$\hat{\sigma}_{SM}(\mathbf{z_i})^{(n)} = \sigma_{SM}(\mathbf{z_i}/T)^{(n)} \tag{5.15}$$

$$\hat{q}_i = \max_n \hat{\sigma}_{SM}(\mathbf{z_i})^{(n)} \tag{5.16}$$

The parameter T is used to control the temperature of a probability distribution. The special case of σ_{SM}, denoted as σ_{SM}, occurs when T is set to 1. By setting T to values greater than 1, we can create a more cautious probability distribution over classes. As T approaches infinity, the probability $\hat{q}_i$ approaches $\frac{1}{N}$, resulting in a uniform distribution and maximum entropy.

Temperature scaling is a crucial step in tuning the empirical hyperparameter T, which is within a narrow range, to achieve optimal model performance [6]. To perform temperature scaling, probability calibration is utilized, which does not require any unknown examples and automatically determines the optimal T through optimization [7].

Probability Calibration

A model is said to be well-calibrated if its predicted confidence scores are close to the true correctness likelihoods. To achieve calibration, we can use temperature scaling to transform the original confidence score $\hat{p}_i$ to a calibrated confidence score $\hat{q}_i$. Given a one-hot representation t and a model prediction y, we can calculate the negative log-likelihood for an example as follows.

$$\mathcal{L} = -\sum_{j=1}^{N} t_j \log y_j \tag{5.17}$$

Then we optimize $\hat{T}$ with respect to negative log-likelihood on the validation set to calibrate the confidence score. We obtain the optimal temperature parameter for SofterMax $\hat{\sigma}_{SM}$ via probability calibration and set T equal to $\hat{T}$ during testing. By applying temperature scaling on softmax, SofterMax retains a relatively more conservative probability distribution for all classes as illustrated in Fig. 5.3. Note that we set T equals to 1 during training. Besides, the probability calibration will not affect the prediction result of known intents.

Decision Boundary

We further tighten the decision boundary of SofterMax outputs by calculating the probability threshold for each class c_i to detect unknown intents. To reduce the open space risk in the probability space, we need to calculate the mean μ_i and standard deviation σ_i of $p(y = c_i \mid x_j, y_j = c_i)$ for each class c_i, where j denotes the jth example. This will give us a probability threshold t_i for each class. To calculate t_i, we use the following formula.

$$t_i = m\ \{0.5, \mu_i - \alpha\sigma_i\} \tag{5.18}$$

The intuition is that we will treat the example, whose probability score is α standard deviations away from mean value as an outlier. For instance, if an example's SofterMax output confidence of each class is lower than the corresponding probability threshold, it will be considered as unknown.

To compare the confidence between different samples, we must calculate a single, comparable confidence score for each sample. We subtract the per-class probability thresholds t_i from the calibrated confidence scores and take the highest value among categories. The lower the confidence score it has, the more likely it is an unknown intent. If the confidence score is lower than 0, we consider the example as unknown. For each sample, we transform the confidence score over classes into a single confidence score as the following:

$$\text{confidence}_{j,i} = p_{j,i} - t_i \qquad (5.19)$$

$$\text{confidence}_j = \max_i \left(\text{confidence}_{j,i} \right) \qquad (5.20)$$

The non-linear nature of the Softmax transformation means that the logits obtained after temperature scaling are not entirely collinear with the original logits. As a result, if we apply the same per-class probability threshold method to calibrate the confidence scores, we may obtain different outcomes when it comes to detecting unknown intents.

To compute the confidence score in the SMDN approach for Softmax, we first apply temperature scaling to the logits, which involves dividing them by a temperature parameter. Next, we subtract the per-class probability thresholds from the scaled logits. Finally, we select the highest value of the resulting scores for a given sample, which serves as the confidence score used for making joint predictions.

5.2.3   Deep Novelty Detection

To detect unfamiliar intentions from various viewpoints, we use a novelty detection algorithm in combination with feature representations that have been learned by deep neural networks.

OpenMax [8] has showcased the potential of reducing open space risk in the feature space. However, our approach diverges from OpenMax by utilizing feature representations extracted from the hidden layer preceding the logits as the feature space, as opposed to the logits themselves. This choice is rooted in prior research [9], which suggests that the feature representations encapsulate more advanced semantic concepts than the logits.

To mitigate the potential risk of encountering unexplored regions in the feature space and to identify novel intents, we utilize the local outlier factor (LOF) method [10], which is a density-based technique for novelty detection. LOF can identify unknown intents within the local area based on the local density. We calculate the LOF score as follows.

$$LOF_k(A) = \frac{\sum\limits_{B \in N_k(A)} \frac{lrd(B)}{lrd(A)}}{|N_k(A)|} \qquad (5.21)$$

$N_k(A)$ refers to the set of k neighboring points to point A in a given dataset. The local reachability density, lrd, is a measure of the density of points around a particular point, taking into account the distance between the neighboring points and the point itself. Specifically, lrd is defined as the inverse of the average reachability distance of the k-nearest neighbors of the point.

$$lrd_k(A) = 1 \bigg/ \left(\frac{\displaystyle\sum_{B \in N_k(A)} reachdist_k(A, B)}{|N_k(A)|} \right) \tag{5.22}$$

The average distance between object A and its neighboring objects can be determined by computing the reachability distance. The reciprocal of this average distance is defined as reachdist$_k(A, B)$.

$$reachdist_k(A, B) = m\{k - \text{distance } (B), d(A, B)\} \tag{5.23}$$

Where the direct distance between A and B is denoted as d(A, B), and the distance of object A to its kth nearest neighbor is referred to as k-distance. If the local density of an example is significantly lower than the local density of its k-nearest neighbor, it is more likely to be classified as anomalous. We consider the LOF score to be a measure of novelty. The higher the LOF score, the greater the likelihood of it being an unknown intention. Furthermore, we believe that other algorithms for detecting novelty can also be applied to this approach.

5.2.4   SMDN

The SMDN approach presented in the study combines the outputs of the SofterMax and deep novelty detection methods to produce a joint prediction. However, as the confidence score computed by SofterMax and the novelty score calculated by LOF are not on the same scale, they cannot be directly integrated. To address this issue, Platt scaling [1] is employed to convert the scores into a novelty probability ranging between 0 and 1, enabling the joint prediction to be made. Platt scaling is a technique that can convert scores to probabilities by training a logistic regression model on the scores. This method is commonly employed in maximum-margin algorithms like SVM to obtain probability estimates.

$$P(y = 1 \mid x) = \frac{1}{1 + \exp(Af(x) + B)} \tag{5.24}$$

Given a function f(x) that represents scores, A and B are scalar parameters learned by the algorithm. The primary objective of Platt scaling is to ensure that samples in the vicinity of the decision boundary have a 50% chance of being identified as unknown (novelty probability), while scaling the scores of the remaining samples to a probability range of 0 to 1. By normalizing the results, we can evaluate the degree of novelty using the same metric for both SMDN and LOF and make a joint prediction, thereby estimating the level of novelty.

5.3 Experiments

The experiment is introduced in six parts, including Datasets, Baseline methods, Experiment settings, Evaluation, Hyper-parameters and experimental results.

5.3.1 Datasets

To examine the resilience and efficacy of the proposed techniques, we performed tests on three openly accessible benchmark conversation datasets, specifically SNIPS, ATIS [11], and SwDA [12], utilizing various experimental configurations. We present comprehensive dataset statistics in Table 5.1.

1. **SNIPS:** We initially perform experiments on the SNIPS personal voice assistant dataset, which includes 7 distinct user intents spanning various domains. The training set comprises 13,084 utterances, while the validation and testing sets each contain 700 utterances. The instances in each class exhibit relatively even distribution.
2. **ATIS** (Airline Travel Information System): The ATIS corpus comprises audio recordings of individuals booking flights and includes 18 distinct user intents pertaining to air travel. The training set contains 4978 utterances, the validation set contains 500 utterances, and the test set contains 893 utterances. The distribution of classes within the ATIS dataset is imbalanced, with the top 25% classes representing approximately 93.7% of the training data.
3. **SwDA** (Switchboard Dialog Act Corpus): SwDA SwDA is a dataset comprising 1155 telephone conversations between two individuals discussing pre-determined topics, with 42 distinct types of speech acts. Given that speech acts can be considered as low-level intent, we seek to ascertain the validity of the current detection approach in this context. However, there is no universally accepted approach for dividing the SwDA dataset into training, validation, and testing sets. In accordance with the data partitioning strategy proposed in [4], we randomly select 80% of conversations for the training set, 10% for the validation set, and 10% for the testing set. The training set consists of 162862 utterances, while the validation and testing sets comprise 20784 and 20146 utterances,

Table 5.1 Statistics of ATIS, SNIPS, and SwDA dataset

Dataset	Classes	Vocabulary	Training	Validation	Test	Turn-taking	Class distribution
SNIPS	7	11,971	13,084	700	700	Single-turn	Balanced
ATIS	18	938	4978	500	893	Single-turn	Imbalanced
SwDA	42	21,812	162,862	20,784	20,146	Multi-turn	Imbalanced

respectively. On average, each conversation comprises 176 sentences. Additionally, the categories in SwDA exhibit a significant imbalance, with the top 25% accounting for around 90.9% of the training data. In Sect. 5.3.3, we will elaborate on how to split existing datasets with various percentages of known intent.

5.3.2   Baselines

We have evaluated our detection techniques by contrasting them with a basic benchmark, the most advanced approach, and its expansion, DOC (Softmax).

1. **Softmax:** ($t = 0.5$) We utilize a confidence threshold on the softmax predictions as a rudimentary benchmark, where the threshold is set to 0.5. To elaborate, if the likelihood of each class output is less than or equal to 0.5, the instance will be classified as unknown.
2. **DOC** [5] is the current state-of-the-art method in the open-world classification problem. The method rejects unknown examples by using the sigmoid activation function as the final layer. It further tightens the decision boundary of the sigmoid function by calculating the confidence threshold for each class through the statistics approach.
3. **DOC (Softmax):** A modified version of DOC that substitutes the sigmoid activation function with softmax.

Please note that we have not contrasted our approach with various other benchmarks cited in [5] as they have already demonstrated that the performance of DOC is considerably superior. Additionally, we have assessed all detection techniques using the same classifier to ensure equitable comparison.

5.3.3   Experiment Settings

We adopt a similar cross-validation methodology as in previous studies [5, 13], where certain classes are treated as unknown during training and mixed back during testing. We vary the proportion of known classes in the training set, specifically using 25%, 50%, and 75% of classes, while using 100% classes during testing. When using all classes during training, this is equivalent to a standard intent classification task. To demonstrate the effectiveness of our classifier architecture in modeling intents, we present the results of using 100% classes during training in Table 5.2.

Table 5.2 Classifier performance of using all classes for training on different datasets (%)

Dataset	Model	Acc*	Acc	Macro F1
SNIPS	BiLSTM [2]	97	97.43	97.47
ATIS	BiLSTM [14]	98.99	98.66	93.99
SwDA	CNN + CNN [4]	78.45	77.44	50.09

* Indicates the performance reported in their original paper

In order to ensure an equitable evaluation of an imbalanced dataset, we employ weighted random sampling without replacement in the training set to randomly select known classes for each experimental run. This method increases the likelihood of selecting classes with a higher number of examples as known classes, while still affording a possibility for classes with fewer examples to be selected with a certain probability. Any other classes that are not chosen as known classes are treated as unknown classes. Please note that we eliminate examples labeled as unknown from both the training and validation sets.

Evaluation

We employ the macro F1-score as the performance measure and assess the outcomes across all categories, recognized categories, and the unidentified category. Our primary attention is on the results pertaining to the unidentified category, aimed at identifying unfamiliar intentions, and we present the average findings over 10 iterations.

Given a collection of categories $C = \{C_1, C_2, C_3, \ldots, C_N\}$, we calculate the macro F1-measure as follows.

$$F_{1,\text{macro}} = 2 \frac{\text{recall}_{\text{macro}} \times \text{precision}_{\text{macro}}}{\text{recall}_{\text{macro}} + \text{precision}_{\text{macro}}} \tag{5.25}$$

$$\text{precision}_{\text{macro}} = \frac{\sum\limits_{i=1}^{N} \text{precision}_{C_i}}{N}, \text{recall}_{\text{macro}} = \frac{\sum\limits_{i=1}^{N} \text{recall}_{C_i}}{N} \tag{5.26}$$

$$\text{precision}_{C_i} = \frac{TP_{C_i}}{TP_{C_i} + FP_{C_i}}, \quad \text{recall}_{C_i} = \frac{TP_{C_i}}{TP_{C_i} + FN_{C_i}} \tag{5.27}$$

Here, C_i represents each individual category within the collection of categories C.

To assess the efficacy of temperature scaling for probability calibration, we employ the Expected Calibration Error [15] (ECE). The primary concept involves partitioning the confidence outputs into K equally sized intervals or "bins" and calculating the weighted mean of the discrepancy between confidence and accuracy across the bins. We compute ECE as follows.

$$ECE = \sum_{i=1}^{K} P(i) * |o_i - e_i| \tag{5.28}$$

Where $P(i)$ signifies the observed probability of all samples that belong to bin i. In this context, o_i represents the actual proportion of positive instances in bin i (accuracy), and e_i denotes the mean calibrated confidence in bin i. The discrepancy between accuracy and confidence is referred to as the confidence gap. A lower value of ECE indicates superior calibration of the model.

Hyper-Parameters

To train classifiers, we commence by initializing the embedding layer with publicly accessible GloVe [16] pre-trained word vectors consisting of 400,000 words and 300 dimensions. Then, we fine-tune the embedding layer by means of back-propagation.". For BiLSTM model, we set the cell output dimension as 128 and dropout rate as 0.5. For both sentence CNN and context CNN in CNN + CNN model, we set the context window size as 3, the kernel size ranging from 1 to 3, and the number of filter maps as 100. We use Adam optimizer with 0.001 learning rate. The maximum training epoch is set as 30 for ATIS and SNIPS and 100 for SwDA. We set the batch size as 128 for SNIPS and ATIS, and 256 for SwDA. We set the fixed input sequence length as their maximum for ATIS and SNIPS. For SwDA, we set the fixed input sequence length as sequence length's mean plus six standard deviations. We implement all models and methods with Keras framework.

For the confidence thresholds of each class in SofterMax and the decision threshold in LOF, we set the α as 2. The intuition is that an example whose confidence score is two standard deviations away from mean will be considered as an outlier (unknown). We set the number of neighbors in LOF as 20 and bin size K in temperature scaling as 15 as suggested by their original implementation.

5.3.4   Experiment Results

In this subsection, we evaluate the results of unknown intent detection on two single-turn dialogue datasets and one multi-turn dialogue dataset. The results are shown in Table 5.3 Besides, to further investigate how will different detection methods affect the existed classification task, we also evaluate the classification performance on the known classes and all classes. The outcomes are presented in Figs. 5.4, 5.5, and 5.6, where the x-axis indicates diverse proportions of categories regarded as recognized intents, and the y-axis represents the macro F1-score. The black error bars depicted in the charts represent a 95% confidence interval. It is recommended to view the figures in color. Please note that the macro F1-scores for the unknown class in

Table 5.3 Macro F1-Score of unknown intent detection on SNIPS, ATIS and SwDA dataset with different proportion of classes treated as known intents

	SNIPS			ATIS			SwDA		
% of known intents	25%	50%	75%	25%	50%	75%	25%	50%	75%
Softmax(t = 0.5)	–	6.15	8.32	8.14	15.3	17.2	19.3	18.4	8.36
DOC	72.5	67.9	63.9	61.6	63.8	37.7	25.4	19.6	7.63
DOC(Softmax)	72.8	65.7	61.8	63.6	63.3	39.7	23.6	18.9	7.67
SofterMax	78.8	70.5	67.2	67.2	65.5	40.7	**28.0**	**20.7**	7.51
LOF	76.0	69.4	65.8	67.3	61.8	38.9	21.1	12.7	4.50
SMDN	**79.8**	**73.0**	**71.0**	**71.1**	**66.6**	**41.7**	20.8	18.4	**8.44**

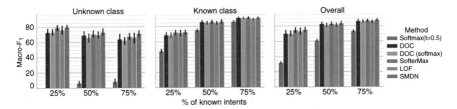

Fig. 5.4 Macro F1-Score on SNIPS dataset with different proportion of classes treated as known intents

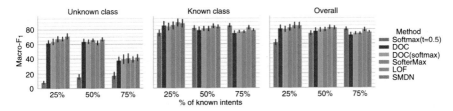

Fig. 5.5 Macro F1-Score on ATIS dataset with different proportion of classes treated as known intents

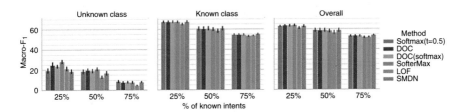

Fig. 5.6 Macro F1-Score on SwDA dataset with different proportion of classes treated as known intents

Figs. 5.4, 5.5, and 5.6 are identical to the scores in Table 5.3. Here we compare its performance with known classes classification and overall classification together.

Single-Turn Dialogue Datasets

Firstly, we assess the outcomes of detecting unidentified intentions on SNIPS and ATIS datasets, as shown in Table 5.3. Comparatively, the SMDN approach surpasses all other techniques in both SNIPS and ATIS datasets. When compared to the state-of-the-art benchmark technique, DOC, SMDN enhances the SNIPS dataset's results by 7.3% in the 25% setting, 5.1% in the 50% setting, and 7.1% in the 75% setting. In the meantime, with respect to SofterMax, it consistently achieves better performance than all baselines across all datasets; with regard to LOF, it outperforms the baselines in SNIPS and performs similarly in ATIS.

For Softmax (with a temperature of 0.5), it does not successfully identify unknown intents in both the SNIPS and ATIS datasets. The performance of DOC and DOC(softmax) is not significantly different. This suggests that the success of DOC is more dependent on the confidence thresholds used in statistical analysis rather than the 1-vs-rest final layer of the sigmoid function. As the number of known intents increases, the error bars in Figs. 5.4 and 5.5 become smaller, indicating greater robustness of the results.

We conduct additional assessments on the effectiveness of the overall categorization, as demonstrated in Figs. 5.4 and 5.5. As we can see, the SofterMax and SMDN method benefit not only the unknown intent detection, but also the overall classification.

Multi-Turn Dialogue Dataset

After we verify the effectiveness of the introduces methods on single-turn dialogue datasets, we evaluate the results of unknown intent detection on SwDA dataset. In Table 5.3 we can see the SofterMax method consistently performs better than all baselines in the 25 % , 50% settings, and the proposed SMDN method manifests slightly better performance in 75% setting. For the results of the overall classification shown in Fig. 5.6 the SofterMax method can benefit the unknown intent detection without degrading the overall classification performance.

We also report the temperature parameter and ECE in Table 5.4 to evaluate the performance of probability calibration. Temperature parameters that are automatically learned by probability calibration are typically in the range of 1.2 to 1.5. After calibration, the ECE has not changed much in both SNIPS and ATIS datasets, and we reduce the ECE by 3% to 4% in SwDA dataset. Notes that we use the validation set which does not contain any unknown intents to calculate ECE. It can only represent how well the model is calibrated with in-domain samples and can not represent the real situation when facing unknown intents. Since we can not merge the probability of unknown intent detection and known intent classification as a whole, we can not calculate the ECE during testing.

Table 5.4 Expected calibration error of the model(ECE)

		SNIPS			ATIS			SwDA	
% of known classes	25%	50%	75%	25%	50%	75%	25%	50%	75%
ECE (uncalibrated)	0.01%	0.11%	0.1%	0.01%	0.5%	0.7%	5.2%	5.9%	7.4%
ECE (temp. Scaling)	0.1%	0.16%	0.1%	0.04%	0.6%	0.8%	2.6%	2.5%	2.8%
Temperature	1.44	1.48	1.28	1.49	1.34	1.36	1.34	1.27	1.33

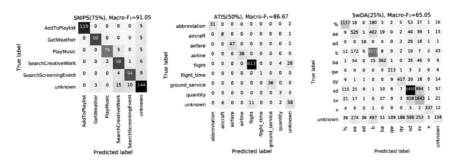

Fig. 5.7 Confusion matrix for the overall classification results with SMDN method on three different datasets

In conclusion, we present the results of our experiments using the confusion matrix illustrated in Fig. 5.7, which illustrates the efficacy of SMDN. Our findings indicate that our approach is proficient in detecting unfamiliar intentions across various dialogue scenarios, as depicted by the majority of classification outcomes aligning with the diagonal.

From the results of different methods, we can observe that merely setting the softmax output confidence threshold as 0.5 does not work well. The DOC method improves performance of unknown intent detection since it calculates the sigmoid output confidence threshold for each known classes based on statistics. However, we can replace the last activation layer of DOC from sigmoid to Softmax and still can get a similar performance. Based on the DOC (Softmax), SofterMax performs probability calibration on the pre-trained classifier through temperature scaling. By calibrating the confidence of the classifier, we can obtain more reasonable confidence thresholds for each class, hence get better performances than baseline methods.

We further detect the unknown intent from different perspectives by feeding feature representations learned by the deep neural network to LOF algorithm. Then, we convert the confidence score of SofterMax and the novelty score of LOF into probabilities through the Platt Scaling. Finally, we obtain the final prediction of SMDN through the joint prediction of SofterMax and LOF. Although the overall performance of LOF is not as good as the SofterMax, we can still greatly improved performance of SMDN through joint prediction.

By taking the different portion of classes as known intents, we can see that when the number of known intents increases, the performance of almost all methods decreases. Taking the results on SNIPS dataset as an example, the macro F1-score of SMDN method drops from 0.798 to 0.71 when the proportion of known intents increases from 25% to 75%. The reason is that when there are more known intents, the semantic meaning of the unknown class is partially overlapping with known classes. Especially in the imbalanced datasets like ATIS and SwDA, their performances drop even more.

We also observe that the results of SwDA dataset are worse than SNIPS and ATIS datasets. It may be due to the data quality of SwDA itself [17]. also point out the

Fig. 5.8 Confidence score
distribution of SofterMax

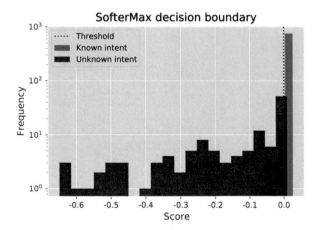

Fig. 5.9 Novelty score
distribution of LOF

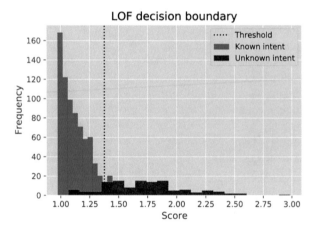

inter-labeler agreement (accuracy) of SwDA is merely 0.84. The noisy labels prevent
the classifier from capturing the high-level semantic concepts of intents, thereby
causing the poor performance on detecting unknown intent. While the base classifier
can achieve 0.974 and 0.986 accuracy on SNIPS and ATIS, respectively, it can only
achieve 0.774 accuracies on SwDA. Still, our method achieves certain improvement
in SwDA compared with baselines.

Finally, we depict the dispersion of assurance and originality using Figs. 5.8 and
5.9 correspondingly. The distribution of scaled novelty probability of SofterMax,
LOF, and SMDN are shown in Fig. 5.10. Notes that the y-axis is in log scale for
better visualization. The green and red bars represent the score distribution of known
and unknown examples, respectively. The vertical dot line indicates the decision
threshold for unknown intent detection. In Fig. 5.9, when utilizing SofterMax, any
example whose confidence score is below the decision threshold will be classified as
unidentified. On the other hand, when using LOF, any example whose novelty score
exceeds the decision threshold will be designated as unidentified. As displayed in

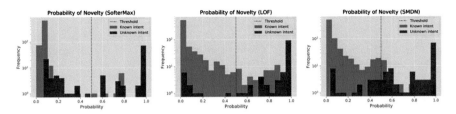

Fig. 5.10 The distribution of novelty probability after Platt scaling for SofterMax, LOF and SMDN

Figs. 5.8 and 5.9, the decision threshold employed by SofterMax and LOF enables the differentiation of unidentified intents from known intents.

For scaled novelty probability distribution in Fig. 5.10, all scores are transformed into the same probability scale ranging from 0 to 1. If the novelty probability of an example is higher than 0.5, it will be considered as unknown. Notes that we only use the per-class probability threshold to calculate the confidence score in Fig. 5.8. Since the novelty probability of SofterMax is derived from the confidence score, it already considers the per-class probability threshold. Therefore, we use 0.5 as the joint prediction threshold, just like the regular binary classification task. We can see that SofterMax treats some examples of unknown intents as the known, while LOF treats some example of known intents as the unknown. After the joint prediction, the probability distribution becomes more than separable, which significantly improves the performance of unknown intent detection.

5.4 Conclusion

In this chapter, we have introduced a simple yet effective post-processing method, SMDN, for detecting unknown intent in the dialogue system. SMDN can be easily applied to a pre-trained deep neural network classifier and requires no changes in the model architecture. The SofterMax calibrates the confidence of Softmax output with temperature scaling to reduce the open space risk in probability space and obtains calibrated decision thresholds for detecting unknown intent. We further combine traditional novelty detection algorithm, LOF, with feature representations learned by the deep neural network. We transform the confidence score of SofterMax and novelty score of LOF into novelty probability to make the joint prediction.

Extensive experiments have been conducted on three benchmark datasets, including two single-turn and one multi-turn dialogue dataset. The results show that our method can yield significant improvements compared with the state-of-the-art baselines. We also believe our method applies to images. For future work, we plan to design a more robust solution that can distinguish unknown intent from known intents even if their semantics meanings are highly similar. Besides, we also plan to use more powerful pre-trained model such as BERT [18] to maximize the benefits of SofterMax.

References

1. Platt, J.: Probabilistic outputs for support vector machines and comparisons to regularized likelihood methods. Adv. Large Mar. Classif. **10**(3), 61–74 (1999)
2. Goo, C., Gao, G., Hsu, Y., et al.: Slot-gated modeling for joint slot filling and intent prediction. Proceedings of the 16th Conference of the North American Chapter of the Association for Computational Linguistics: Human Language Technologies, pp. 753–757 (2018)
3. Graves, A., Fernández, S., Schmidhuber, J.: Bidirectional LSTM networks for improved phoneme classification and recognition. Proceedings of the 15th International Conference on Artificial Neural Networks, pp. 799–804 (2005)
4. Liu, Y., Han, K., Tan, Z., et al.: Using context information for dialog act classification in DNN framework. Proceedings of the 22nd Conference on Empirical Methods in Natural Language Processing, pp. 2170–2178 (2017)
5. Shu, L., Xu, H., Liu, B.: DOC: deep open classification of text documents. Proceedings of the 22nd Conference on Empirical Methods in Natural Language Processing, pp. 2911–2916 (2017)
6. Hinton, G.E., Vinyals, O., Dean, J.: Distilling the knowledge in a neural network. Stat (2015)
7. Guo, C., Pleiss, G., Sun, Y., et al.: On calibration of modern neural networks. Proceedings of the 34th International Conference on Machine Learning, pp. 1321–1330 (2017)
8. Bendale, A., Boult, T.E.: Towards open set deep networks. Proceedings of the 39th Institute of Electrical an Electronics Engineers Conference on Computer Vision and Pattern Recognition, pp. 1563–1572 (2016)
9. Ouyang, W., Wang, X., Zeng, X., et al.: Deepid-net: deformable deep convolutional neural networks for object detection. Proceedings of the 39th Institute of Electrical and Electronics Engineers Conference on Computer Vision and Pattern Recognition, pp. 2403–2412 (2015)
10. Breunig, M., Kriegel, H.-P., Ng, R.T., et al.: LOF: identifying density-based local outliers. Proceedings of the 29th Special Interest Group on Management of Data international conference on Management of data, pp. 93–104 (2000)
11. Hemphill, C.T., Godfrey, J.J., Doddington, G.R., et al.: The atis spoken language systems pilot corpus. Proceedings of the Speech and Natural Language, pp. 96–101 (1990)
12. Jurafsky, D.: Switchboard discourse language modeling project final report. Proceedings of the 4th Johns Hopkins Large Vocabulary Continuous Speech Recognition Workshop (1998)
13. Fei G, Liu B. Breaking the closed world assumption in text classification[C]. Proceedings of the 14th Conference of the North American Chapter of the Association for Computational Linguistics: Human Language Technologies, pp. 506–514 (2016)
14. Wang, Y., Shen, Y., Jin, H.: A bi-model based RNN semantic frame parsing model for intent detection and slot filling. Proceedings of the 16th Conference of the North American Chapter of the Association for Computational Linguistics: Human Language Technologies, pp. 309–314 (2018)
15. DeGroot, M.H., Fienberg, S.E.: The comparison and evaluation of forecasters. J. R. Statist. Soc. Ser.D. **32**(1–2), 12–22 (1983)
16. Pennington, J., Socher, R., Manning, C.: Glove: global vectors for word representation. Proceedings of the 19th Conference on Empirical Methods in Natural Language Processing, pp. 1532–1543 (2014)
17. Lee, J.Y., Dernoncourt, F.: Sequential short-text classification with recurrent and convolutional neural networks. Proceedings of the 14th Conference of the North American Chapter of the Association for Computational Linguistics: Human Language Technologies, pp. 515–520 (2016)
18. Devlin J, Chang M-W, Lee K, et al. BERT: Pre-training of deep bidirectional transformers for language understanding[C]. Proceedings of the 17th Conference of the North American Chapter of the Association for Computational Linguistics: Human Language Technologies, pp. 4171–4186 (2019)

Chapter 6
Unknown Intent Detection Based on Large-Margin Cosine Loss

Abstract Identifying the unknown (novel) user intents that have never appeared in the training set is a challenging task in the dialogue system. This chapter presents a two-stage method for detecting unknown intents. The method use a bidirectional long short-term memory (BiLSTM) network with the margin loss as the feature extractor. With margin loss, the network is encouraged to maximize inter-class variance and minimize intra-class variance, thereby learning descriminative deep feature. Then, the feature vectors are fed into a density-based novelty detection algorithm, local outlier factor (LOF), to detect unknown intents. Experiments on two benchmark datasets show that the proposed method can yield consistent improvements compared with the baseline methods.

Keywords Dialogue system · Two-stage method · Margin loss · Deep features · Local outlier factor

6.1 Introduction

In Chap. 5, a new intent detection method based on model post-processing is introduced, which can equip deep neural network classifiers with the ability to detect new intents without affecting the original model architecture. Compared with traditional methods, the key to its performance improvement is that the neural network classifier can more accurately model the intent and obtain a deep intent representation. However, the deep intent representation learned through the traditional Softmax cross-entropy loss function is just a by-product of the neural network model, not specifically designed for new intent detection tasks, and there is still room for improvement.

In this chapter, we will further study the loss function of the deep neural network classifier, so that the model can learn a deep intent representation that is more suitable for new intent detection tasks. Therefore, a large marginal cosine loss function is introduced to replace the Softmax cross-entropy loss function in the traditional classification model. The model must also consider the marginal item penalty while learning the classification decision boundary to obtain the intra-class

H. Xu et al., *Intent Recognition for Human-Machine Interactions*, SpringerBriefs in Computer Science, https://doi.org/10.1007/978-981-99-3885-8_6

compactness and inter-class separation intent representation, thereby improving new intent detection performance.

6.2   New Intent Detection Model Based on Large Margin Cosine Loss Function

This section will provide a detailed explanation of the deep unknown intent detection model using margin loss. First, we presents a two-stage method for unknown intent detection with BiLSTM. Second, we introduce margin loss on BiLSTM to learn discriminative deep features, which is suitable for the detection task. Finally, experiments conducted on two benchmark dialogue datasets show the effectiveness of the proposed method.

To begin with, the method use BiLSTM [1] to train the intent classifier and use it as feature extractor. Figure 6.1 shows the architecture of the proposed method. Given an utterance with maximum word sequence length ℓ, the input word sequence $w_{1 \, : \, \ell}$ is transformed into m-dimensional word embedding $v_{1 \, : \, \ell}$, which is used by forward and backward LSTM to produce feature representations x:

$$\overrightarrow{x_t} = LSTM\left(v_t, \overrightarrow{c_{t-1}}\right) \tag{6.1}$$

$$\overleftarrow{x_t} = LSTM\left(v_t, \overleftarrow{c_{t+1}}\right) \tag{6.2}$$

$$x = \left[\overrightarrow{x_\ell}; \overleftarrow{x_1}\right] \tag{6.3}$$

In the given paragraph, v_t refers to the word embedding of the input at time step t. $\overrightarrow{x_t}$ and $\overleftarrow{x_t}$ are the forward and backward LSTM output vectors, respectively. $\overrightarrow{c_t}$ and $\overleftarrow{c_t}$ are the cell state vectors of the forward and backward LSTM, respectively. To obtain

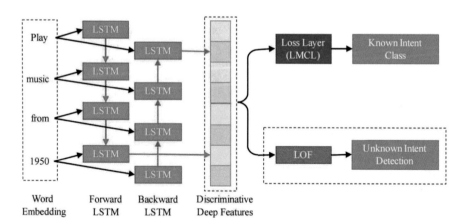

Fig. 6.1   A new intent detection model based on large marginal cosine loss function

the sentence representation x, the last output vector of the forward LSTM $\overrightarrow{x_\ell}$ is concatenated with the first output vector of the backward LSTM $\overleftarrow{x_1}$. This captures the high-level semantic concepts learned by the model, and x is used as input for the next stage.

6.2.1 Large Margin Cosine Loss (LMCL)

Our definition for $\mathcal{L}_{LMC}$ (LMCL) [2] is presented below.

$$\mathcal{L}_{LMC} = \frac{1}{N} \sum_{i}^{N} - \log \frac{e^{s \cdot \left(\cos\left(\theta_{y_i,i}\right) - m\right)}}{e^{s \cdot \left(\cos\left(\theta_{y_i,i}\right) - m\right)} + \sum_{j \neq y_i} e^{s \cdot \cos\left(\theta_{j,i}\right)}} \tag{6.4}$$

constrained by:

$$\cos\left(\theta_j, i\right) = W_j^T x_i \tag{6.5}$$

$$W = \frac{W^\star}{\| W^\star \|} \tag{6.6}$$

$$x = \frac{x^\star}{\| x^\star \|} \tag{6.7}$$

where N denotes the number of training samples, y_i is the ground-truth class of the i-th sample, s is the scaling factor, m is the cosine margin, W_j is the weight vector of the j-th class, and θ_j is the angle between W_j and x_i.

LMCL transforms Softmax loss into cosine loss by applying L2 normalization on both features and weight vectors. It further maximizes the decision margin in the angular space. With normalization and cosine margin, LMCL forces the model to maximize inter-class variance and to minimize intra-class variance. Then, the model is utilized as the feature extractor to produce discriminative intent representations.

6.3 Experiments

6.3.1 Datasets

We have performed tests on two openly accessible standard dialogue datasets, specifically SNIPS and ATIS [3].

6.3.2   Baselines

We adopt the validation methodology described in [4, 5], where certain categories are designated as unknown during training but later incorporated during testing. Then we vary the number of known classes in training set in the range of 25%, 50%, and 75% classes and use all classes for testing.

To conduct a fair evaluation for the imbalanced dataset, we randomly select known classes by weighted random sampling without replacement in the training set. If a class has more examples, it is more likely to be chosen as the known class. Meanwhile, the class with fewer examples still have a chance to be selected. Other classes are regarded as unknown and we will remove them in the training and validation set.

The techniques are evaluated against the most advanced approaches and a modified version of the proposed technique.

1. **Maximum Softmax Probability (MSP)** [6]: Consider the highest possible probability obtained through Softmax as the score for a particular sample. If a sample does not pertain to any recognized intents, its score will be relatively lower. To establish a rudimentary baseline, we determine and apply a confidence threshold on the score, which is established at 0.5.
2. **DOC** [5]: It is the state-of-the-art method in the field of open-world classification. It replaces Softmax with sigmoid activation function as the final layer. It further tightens the decision boundary of the sigmoid function by calculating the confidence threshold for each class through statistics approach.
3. **DOC (Softmax)**: A variant of DOC replaces the sigmoid activation function with Softmax.
4. **LOF (Softmax)**: A variant of the introduces method for ablation study. Softmax loss is employed to train the feature extractor rather than LMCL.

6.3.3   Experiment Setting

Hyper-Parameter Setting

The embedding layer is initialized using pre-trained word vectors from GloVe [7]. The BiLSTM model is configured with an output dimension of 128 and a maximum of 200 epochs, with early stopping enabled. The LMCL and LOF models are set up according to the original research. The evaluation metric used is the macro F1-score, and results are averaged over 10 runs. The scaling factors are set to 30, and the cosine margin is set to 0.35, as recommended by [8].

6.3.4   Experiment Results

We show the experiment results in Table 6.1. Firstly, our method consistently performs better than all baselines in all settings. Compared with DOC, our method improves the macro f1-score on SNIPS by 6.7%, 16.2% and 14.9% in 25%, 50%, and 75% setting respectively. It confirms the effectiveness of our two-stage approach.

Additionally, the approach outperforms LOF (Softmax) as well. As shown in Fig. 6.2, t-SNE [9] is utilized to visualize the deep features that were acquired through the use of Softmax and LMCL. We can see that the deep features learned with LMCL are intra-class compact and inter-class separable, which is beneficial for novelty detection algorithms based on local density.

Thirdly, we observe that on the ATIS dataset, the performance of unknown intent detection dramatically drops as the known intent increases. We think the reason is that the intents of ATIS are all in the same domain and they are very similar in semantics (e.g., flight and flight_no). The semantics of the unknown intents can easily overlap with the known intents, which leads to the poor performance of all methods.

Finally, compared with ATIS, our approach has improved even better on SNIPS. Since the intent of SNIPS is originated from different domains, it causes the DNN to

Table 6.1 Macro f1-score of unknown intent detection with different proportion (25%, 50% and 75%) of classes are treated as known intents on SNIPS and ATIS dataset

	SNIPS			ATIS		
% of known intends	25%	50%	75%	25%	50%	75%
MSP	0.0	6.2	8.3	8.1	15.3	17.2
DOC	72.5	67.9	63.9	61.6	62.8	37.7
DOC (Softmax)	72.8	65.7	61.8	63.6	63.3	38.7
LOF (Softmax)	76.0	69.4	65.8	67.3	61.8	38.9
LOF (LMCL)	**79.2**	**84.1**	**78.8**	**69.6**	**63.4**	**39.6**

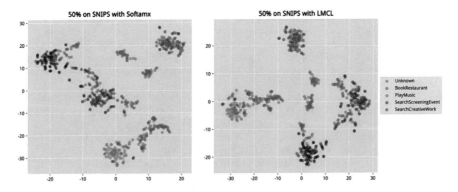

Fig. 6.2 Visualization of deep features learned with Softmax and LMCL on SNIPS dataset

learn a simple decision function when the known intents are dissimilar to each other. By replacing the Softmax loss with the margin loss, the network can be further pushed to reduce the intra-class variance and the inter-class variance, thus improving the robustness of the feature extractor.

6.4 Conclusion

In this chapter, a two-stage method for unknown intent detection has been presented. Firstly, a BiLSTM classifier is trained as the feature extractor. Secondly, Softmax loss is replaced with margin loss to learn discriminative deep features by forcing the network to maximize inter-class variance and to minimize intra-class variance. Finally, unknown intents are detected through the novelty detection algorithm. It is also believe that broader families of anomaly detection algorithms are also applicable to our method.

Extensive experiments conducted on two benchmark datasets show that our method can yield consistent improvements compared with the baseline methods. In future work, the plan is to design a solution that can identify the unknown intent from known intents and cluster the unknown intents in an end-to-end fashion.

References

1. Mesnil, G., Dauphin, Y., Yao, K., et al.: Using recurrent neural networks for slot filling in spoken language understanding. IEEE/ACM Trans. Audio Speech Lang. Process. 23(3), 530–539 (2014)
2. Wang, H., Wang, Y., Zhou, Z., et al.: Cosface: large margin cosine loss for deep face recognition. Proceedings of the 41st Institute of Electrical and Electronics Engineers Conference on Computer Vision and Pattern Recognition, pp. 5265–5274 (2018)
3. Tur, G., Hakkani-Tür, D., Heck L.: What is left to be understood in atis? IEEE Spoken Language Technology Workshop, pp. 19–24 (2010)
4. Fei, G., Liu, B.: Breaking the closed world assumption in text classification. Proceedings of the 14th Conference of the North American Chapter of the Association for Computational Linguistics: Human Language Technologies, pp. 506–514 (2016)
5. Shu, L., Xu, H., Liu, B.: DOC: deep open classification of text documents. Proceedings of the 22nd Conference on Empirical Methods in Natural Language Processing, pp. 2911–2916 (2017)
6. Hendrycks, D., Gimpel, K.: A baseline for detecting misclassified and out-of-distribution examples in neural networks. Proceedings of the 5th International Conference on Learning Representations (2017)
7. Pennington, J., Socher, R., Manning, C.: Glove: Global vectors for word representation. Proceedings of the 22nd Conference on Empirical Methods in Natural Language Processing, pp. 1532–1543 (2014)
8. Wang, F., Cheng, J., Liu, W., et al.: Additive margin softmax for face verification. Inst. Electr. Electron. Eng. Signal Process. Lett. 25(7), 926–930 (2018)
9. Maaten, L.V.D., Hinton, G.: Visualizing data using t-SNE. J. Mach. Learn. Res. 9, 2579–2605 (2008)

Chapter 7
Unknown Intention Detection Method Based on Dynamic Constraint Boundary

Abstract Open intent classification is a challenging task in dialogue systems. On the one hand, ensuring the classification quality of known intents is crucial. On the other hand, identifying open (unknown) intent during testing. Current models are limited in finding the appropriate decision boundary to balance the performances of both known and open intents. In this chapter, a post-processing method is proposed to learn the adaptive decision boundary (ADB) for open intent classification. The model is initially pre-trained using labeled known intent samples. Then, well-trained features to automatically learn the adaptive spherical decision boundaries for each known intent. Specifically, a new loss function is introduced to balance both the empirical risk and the open space risk. This method does not need open samples and is free from modifying the model architecture. This approach is surprisingly insensitive with less labeled data and fewer known intents. Extensive experiments conducted on three benchmark datasets demonstrate significant improvements compared to state-of-the-art methods.

Keywords Open intent classification · Post-processing · Adaptive decision boundary · Empirical risk · Open space risk

7.1 Introduction

Identifying the user's open intent plays a significant role in dialogue systems. Assuming there are two known intents for specific purposes, such as book flight and restaurant reservation. However, there are also utterances with irrelevant or unsupported intents that our system cannot handle. It is necessary to distinguish these utterances from the known intents as much as possible. On the one hand, effectively identifying the open intent can improve customer satisfaction by reducing false-positive error. On the other hand, the open intent can be utilized to discover potential user needs.

In this chapter, a post-processing method to learn the adaptive decision boundary (ADB) for open intent classification is introduced. The suitable decision boundaries should satisfy two conditions. On the one hand, they should be broad enough to surround in-domain samples as much as possible. On the other hand, they need to be

© The Author(s), under exclusive license to Springer Nature Singapore Pte Ltd. 2023
H. Xu et al., *Intent Recognition for Human-Machine Interactions*, SpringerBriefs in Computer Science, https://doi.org/10.1007/978-981-99-3885-8_7

tight enough to prevent out-of-domain samples from being identified as in-domain samples. To address these issues, we introduce a new loss function, which optimizes the boundary parameters by balancing both the open space risk and the empirical risk. The decision boundaries can automatically learn to adapt to the intent feature space until balance with the boundary loss. It is found that the post-processing method can still learn discriminative decision boundaries to detect the open intent even without modifying the original model architecture.

7.2   The Frame Structure of the Model

To perform open intention classification, established intentions are leveraged as prior knowledge and propose a novel post-processing method to learn the adaptive decision boundary (ADB), as depicted in Fig. 7.1. Initially, intention representations are extracted from the BERT model and pre-train the model using the Soft-max loss as a supervisory signal. Centroids are established for each known class and restrict the features of known intents within closed ball regions. Subsequently, our goal is to determine the radius of each ball area to acquire the decision boundaries. The boundary parameters are initialized with a standard normal distribution and employ a learnable activation function as a projection to obtain the radius of each decision boundary.

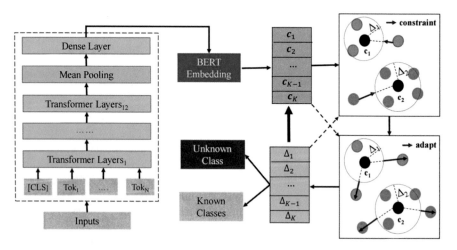

Fig. 7.1 The model architecture of dialogue intention discovery model based on dynamic constraint boundary

7.3 The Main Approach

7.3.1 Intent Representation

The BERT model is utilized for extracting profound intention characteristics. After feeding the ith input sentence s_i into BERT, all token embeddings $[C, T_1, T_2, \ldots, T_N] \in R^{(N + 1) \times H}$ are obtained from the final hidden layer. Following the recommendation in [1], mean-pooling is applied to these token embeddings to amalgamate the high-level semantic traits in a sentence, thereby obtaining the averaged representation $x_i \in R^H$.

$$x_i = \text{mean} - \text{pooling}([C, T_1, T_2, \ldots, T_N]) \tag{7.1}$$

To enhance the feature extraction ability, a dense layer h is utilized to process x_i and obtain the intent representation $z_i \in R^D$, where C represents the vector for text classification, N represents the sequence length, and H represents the hidden layer size. The intention is to strengthen the feature extraction capability by feeding x_i into a dense layer h and obtaining the intent representation $z_i \in R^D$. The sequence length is denoted by N, and the hidden layer size is denoted by H.

$$z_i = \sigma(W_h x_i + b_h) \tag{7.2}$$

The dimension of the intent representation is denoted as D. The ReLU activation function is represented by σ. The layer h has weight and bias terms denoted as W_h and b_h, respectively. Specifically, W_h is a matrix with dimensions of H by D and b_h is a vector with dimensions of D.

7.3.2 Pre-Training

In order to enable the learned decision boundaries to adjust to the intent feature space, it is necessary to first acquire representations of intent. Given the scarcity of available intent samples, existing intents are employed as prior knowledge to pre-train the model. To assess the efficacy of the learned decision boundary, a straightforward Softmax loss function L_s are employed to learn the intent feature z_i.

$$L_s = -\frac{1}{N} \sum_{i=1}^{N} -\log \frac{exp(\varnothing(z_i)^{y_i})}{\sum_{j=1}^{K} exp(\varnothing(z_i)^j)} \tag{7.3}$$

A linear classifier denoted by $\varnothing(\cdot)$ and its corresponding output logits $\varnothing(\cdot)^j$ for the jth class are utilized. Afterwards, the pre-trained model is employed to extract intent features which will be used in learning the decision boundary.

7.3.3 Adaptive Decision Boundary Learning

In this section, an approach is suggested to acquire the adaptive decision boundary (ADB) for open intent classification. Initially, the definition of the decision boundary is presented. Subsequently, a strategy for boundary learning is proposed to achieve optimization. Lastly, the acquired decision boundary is employed for open classification.

Decision Boundary Formulation

Previous research [2] has demonstrated the effectiveness of using a spherical-shaped boundary for open classification. In contrast to the half-space binary linear classifier [3] or two parallel hyper-planes [4], a bounded spherical area can significantly mitigate the risk associated with open space. Motivated by this observation, the endeavor is made to learn the decision boundary of each class such that the known intents are restricted within a spherical area.

$$c_K = \frac{1}{|s_k|} \sum_{(z_i, y_i) \in s_i}^{N} z_i \tag{7.4}$$

Given a set of examples s_k, where $|s_k|$ represents the number of examples in s_k, Δ_k can be defined as the distance between the decision boundary and the centroid c_K. The objective is to satisfy the following constraints for each known intent z_i.

$$\forall z_i \in s_k, \|z_i - c_k\|_2 \leq \Delta_k \tag{7.5}$$

The aim is to confine examples belonging to class k within a ball area centered at c_k with a radius of Δ_k, where $\|z_i - c_k\|_2$ represents the Euclidean distance between z_i and c_k. To make the radius Δ_k adaptable to different feature spaces, a deep neural network is employed to optimize the learnable boundary parameter $\widehat{\Delta_k} \in R$. Following the recommendation in [5], the Softplus activation function is applied as a mapping between Δ_k and $\widehat{\Delta_k}$:

$$\Delta_k = \log\left(1 + e^{\widehat{\Delta_k}}\right) \tag{7.6}$$

The Softplus activation function offers several benefits. Firstly, it is fully differentiable for any value of $\widehat{\Delta_k} \in R$. Secondly, it guarantees a positive learned radius Δ_k. Lastly, it exhibits linear behavior similar to ReLU and can accommodate larger values of Δ_k if needed.

Boundary Learning

The adaptability of decision boundaries to the intent feature space is crucial for balancing empirical and open space risk [6]. For instance, when $\|z_i - c_k\|_2 > \Delta_k$, the known intent samples fall outside their corresponding decision boundaries, which may increase empirical risk. Therefore, decision boundaries should expand to encompass more samples from known classes. On the other hand, if $\|z_i - c_k\|_2 < \Delta_k$, broader decision boundaries may result in the identification of more known intent samples, but they could also introduce more open intent samples and raise open space risk. Therefore, we introduce the boundary loss L_b.

$$L_b = \frac{1}{N} \sum_{i=1}^{N} \left[\delta_i \left(\|z_i - c_{y_i}\|_2 - \Delta_{y_i} \right) + (1 - \delta_i) \left(\Delta_{y_i} - \|z_i - c_{y_i}\|_2 \right) \right] \tag{7.7}$$

The label of the ith sample can be expressed as y_i, and define δ_i as follows.

$$\delta_i := \begin{cases} 1, & \text{if } \|z_i - c_{y_i}\|_2 > \Delta_k \\ 0, & \text{if } \|z_i - c_{y_i}\|_2 < \Delta_k \end{cases} \tag{7.8}$$

To update the $\widehat{\Delta_k}$ value in relation to L_b, the following modifications can be implemented.

$$\widehat{\Delta_k} := \widehat{\Delta_k} - \eta \frac{\partial L_b}{\partial \widehat{\Delta_k}} \tag{7.9}$$

The learning rate for the boundary parameters $\widehat{\Delta_k}$ is denoted by η, and $\frac{\partial L_b}{\partial \widehat{\Delta_k}}$ is calculated by:

$$\frac{\partial L_b}{\partial \widehat{\Delta}_k} = \frac{\sum_{i=1}^{N} \delta_i'(y_i = k) \cdot (-1)^{\delta_i}}{\sum_{i=1}^{N} \delta_i'(y_i = k)} \cdot \frac{1}{1 + e^{-\widehat{\Delta}_k}} \tag{7.10}$$

where $\delta_i'(y_i = k) = 1$ if $y_i = k$ and $\delta_i'(y_i = k) = 0$ if not. To avoid division by zero, only the radius Δ_{y_i} corresponding to class k is adjusted in a mini-batch update.

Using the boundary loss, denoted as L_b, enables the adaptation of boundaries to the intent feature space, facilitating the learning of appropriate decision boundaries. These learned decision boundaries are effective in encircling most of the known intent samples and are also not situated far from each known class centroid. This efficacy allows for the identification of open intent samples. Hence, it is a valuable technique for classifying various intents.

7.3.4 Open Classification with Decision Boundary

Upon completion of training, the acquired cluster centroids and decision boundaries are utilized for inference. It is assume that the acknowledged intent samples are confined within the enclosed spherical region generated by their respective centroids and decision boundaries. Conversely, the uncovered intent samples exist outside all the constrained spherical regions. To be more precise, open intent classification is executed in the following manner:

$$\widehat{y} = \begin{cases} \text{open, if } d(z_i, c_k) > \Delta_k, \forall k \in y \\ \text{argmin}_{k \in y} d(z_i, c_k), \text{otherwise} \end{cases} \tag{7.11}$$

where $d(z_i, c_k)$ denotes the Euclidean distance between z_i and c_k denote the known intent labels.

7.4 Experiments

7.4.1 Datasets

We perform tests on three demanding actual datasets to assess our method. The comprehensive figures are presented in Table 7.1.

1. **BANKING:** A fine-grained dataset in a banking domain [7]. It contains 77 intents and 13,083 customer service queries.
2. **OOS:** A dataset for intent classification and out-of-scope prediction [8]. It contains 150 intents, 22,500 in-domain queries and 1200 out-of-domain queries.

Table 7.1 Statistics of BANKING, OOS and StackOverflow datasets. (# indicates the total number of sentences)

Dataset	Classes	#Training	#Validation	#Test	Vocabulary Size	Length (max / mean)
BANKING	77	9003	1000	3080	5028	79/11.91
StackOverflow	20	12,000	2000	6000	17,182	41/9.18
OOS	150	15,000	3000	5700	7447	28/18.42

3. **StackOverflow:** A dataset published in Kaggle.com. It contains 3,370,528 technical question titles. The processed dataset [9] is utilized, which includes 20 different classes and 1,000 samples for each class.

7.4.2 Baselines

The approach is evaluated against the leading open classification techniques: OpenMax [10], MSP [11], DOC [12], and DeepUnk [13], which represent the current state-of-the-art in this field.

The OpenMax approach, originally used in computer vision for open set detection, is employed for open intent classification. Initially, the Softmax loss is utilized to train a classifier on known intents, followed by fitting a Weibull distribution to the output logits of the classifier. Ultimately, the confidence scores are recalibrated using the OpenMax Layer. Since there were enough open intents for fine-tuning, the default hyper-parameters of OpenMax are utilized. To ensure a fair comparison, the backbone network of other methods is substituted with the same BERT model used in our study. For the MSP, the identical confidence threshold (0.5) presented in [13] is used.

7.4.3 Experiment Settings

Using the same approach as described in [12, 13], we maintain certain classes as unknown (open) and reintegrate them during the testing phase. All datasets are partitioned into training, validation, and testing sets. Initially, we manipulate the number of known classes by adjusting their proportions to 25%, 50%, and 75% in the training set. Afterwards, we consider the remaining classes as open classes and exclude them from the training set. Finally, we incorporate both the known classes and open classes for testing. For each known class ratio, we calculate the average performance by conducting ten experiments.

Evaluation Metrics

We adhere to the approach of previous studies [12, 13] and designate all classes, except for the known classes, as a rejected open class. Accuracy score (Accuracy) and macro F1-score (F1-score) are utilized as metrics to evaluate the overall performance, considering all classes (both known and open classes). Moreover, we measure macro F1-score individually over the known classes and open class to better evaluate the performance at a more detailed level.

7.4.4 Experiment Results

Tables 7.2 and 7.3 show the performances of all compared methods, where the best results are highlighted in bold. Firstly, we observe the overall performance. Table 7.2 shows accuracy score and macro F1-score over all classes. "Accuracy" and "F1-score" represent the precise measurement and macro F1-score across all categories. Our methodology consistently exhibits the most exceptional results and surpasses other benchmarks by a substantial margin, with 25%, 50%, and 75% known categories. When compared to all the baselines' best results, our approach enhances the accuracy score (Accuracy) by 14.64%, 6.13%, and 2.56% in BANKING, 6.16%, 3.19%, and 2.61% in OOS, and 38.88%, 27.42%, and 10.45% in StackOverflow, respectively, in the 25%, 50%, and 75% configurations, which highlights the superiority of our technique.

Table 7.2 Results of open classification with different known class proportions(25%, 50% and 75%) on BANKING, OOS and StackOverflflow datasets

		BANKING		OOS		StackOverflow	
	Methods	Accuracy	F1-score	Accuracy	F1-score	Accuracy	F1-score
25%	MSP	43.67	50.09	47.02	47.62	28.67	37.85
	DOC	56.99	58.03	74.97	66.37	42.74	47.73
	OpenMax	49.94	54.14	68.50	61.99	40.28	45.98
	DeepUnk	64.21	61.36	81.43	71.16	47.84	52.05
	ADB	**78.85**	**71.62**	**87.59**	**77.19**	**86.72**	**80.83**
50%	MSP	59.73	71.18	62.96	70.41	52.42	63.01
	DOC	64.81	73.12	77.16	78.26	52.53	62.84
	OpenMax	65.31	74.24	80.11	80.56	60.35	68.18
	DeepUnk	72.73	77.53	83.35	82.16	58.98	68.01
	ADB	**78.86**	**80.90**	**86.54**	**85.05**	**86.40**	**85.83**
75%	MSP	75.89	83.60	74.07	82.38	72.17	77.95
	DOC	76.77	83.34	78.73	83.59	68.91	75.06
	OpenMax	77.45	84.07	76.80	73.16	74.42	79.78
	DeepUnk	78.52	84.31	83.71	86.23	72.33	78.28
	ADB	**81.08**	**85.96**	**86.32**	**88.53**	**82.78**	**85.99**

Table 7.3 Results of open classification with different known class ratios(25%, 50% and 75%) on BANKING, OOS and StackOverflflow datasets

	Methods	BANKING		OOS		StackOverflow	
		Open	Known	Open	Known	Open	Known
25%	MSP	41.43	50.55	50.88	47.53	13.03	42.82
	DOC	61.42	57.85	81.98	65.96	41.25	49.02
	OpenMax	51.32	54.28	75.76	61.62	36.41	47.89
	DeepUnk	70.44	60.88	87.33	70.73	49.29	52.60
	ADB	**84.56**	**70.94**	**91.84**	**76.80**	**90.88**	**78.82**
50%	MSP	41.19	71.97	57.62	70.58	23.99	66.91
	DOC	55.14	73.59	79.00	78.25	25.44	66.58
	OpenMax	54.33	74.76	81.89	80.54	45.00	70.49
	DeepUnk	69.53	77.74	85.85	82.11	43.01	70.51
	ADB	**78.44**	**80.96**	**88.65**	**85.00**	**87.34**	**85.68**
75%	MSP	39.23	84.36	59.08	82.59	33.96	80.88
	DOC	50.60	83.91	72.87	83.69	16.76	78.95
	OpenMax	50.85	84.64	76.35	73.13	44.87	82.11
	DeepUnk	58.54	84.75	81.15	86.27	37.59	81.00
	ADB	**66.47**	**86.29**	**83.92**	**88.58**	**73.86**	**86.80**

We have observed that the improvements on StackOverflow are considerably more significant than the other two datasets. We believe that this is mainly due to the characteristics of the datasets. Most of the baseline methods lack explicit or suitable decision boundaries to accurately identify the open intent, making them more sensitive to variations in different datasets. However, for StackOverflow, these methods are limited in their ability to differentiate complex semantic intents (such as technical question titles) without prior knowledge. On the other hand, our approach is able to learn specific and well-defined decision boundaries for each known class, making it more effective for open intent classification.

We have also observed the fine-grained performance, as shown in Table 7.3, which presents the macro F1-score for open intent and known intents, respectively. The "Open" and "Known" categories refer to the macro F1-score over the open class and known classes, respectively. We have noticed that our method not only achieves significant improvements in open class classification but also greatly enhances the performance on known classes compared to the baseline methods. This can be attributed to the fact that our approach is capable of learning specific and well-defined decision boundaries for detecting open class while ensuring high-quality classification of known intents.

7.5 Discussion

7.5.1 Boundary Learning Process

Figure 7.2 depicts the process of learning the decision boundary. Initially, the majority of the parameters are initialized with small values near zero, resulting in a small radius with the Softplus activation function. Subsequently, due to the initial small radius, the empirical risk dominates the process, leading to an expansion of the radius of each decision boundary to encompass more known intent samples that belong to its class. During training, the radius of the decision boundary progressively grows to enclose most of the known intents, while also introducing redundant open intent samples with a large radius. At this stage, the open space risk becomes predominant, thereby inhibiting further radius expansion. Eventually, the decision boundaries converge, balancing the empirical risk and open space risk.

7.5.2 Effect of Decision Boundary

In order to assess the efficacy of the learned decision boundary, different ratios of Δ are utilized as boundaries during testing. Figure 7.3 illustrates that ADB outperforms all other decision boundaries, which attests to the precision of the learned decision boundary. Additionally, we observed that the effectiveness of open classification is highly dependent on the accuracy of the decision boundaries. Exceeding compact boundaries increases the risk of misclassifying known intent samples as open intent, which would elevate the open space risk. Conversely, loosened boundaries increase the empirical risk by misidentifying open intent samples as known intents. As demonstrated in Fig. 7.3, both of these scenarios underperform when compared to Δ.

Fig. 7.2 The boundary learning process

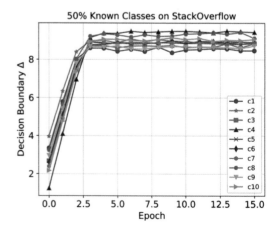

Fig. 7.3 Influence of the learned decision boundary

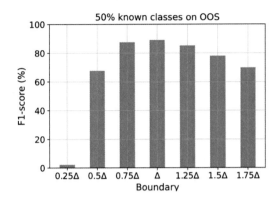

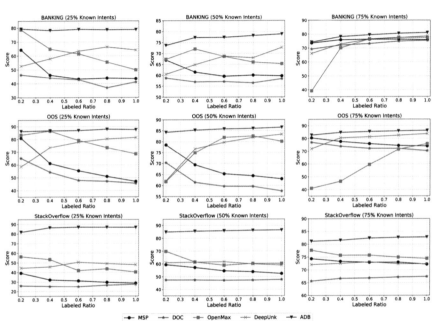

Fig. 7.4 Influence of labeled ratio on three datasets with different known class proportions (25%, 50%, 75%)

7.5.3 *Effect of Labeled Data*

To investigate the influence of the labeled ratio, the labeled data in the training set is varied in the range of 0.2, 0.4, 0.6, 0.8 and 1.0. Accuracy is used as the score to evaluate the performance. As shown in Fig. 7.4, we find ADB outperforms all the other baselines on three datasets on almost all settings. Besides, ADB keeps a more robust performance under different labeled ratios compared with other methods.

We have observed that statistical methods such as MSP and DOC exhibit superior performance when working with smaller amounts of labeled data. This may be attributed to their ability to generate lower predicted confidence scores with limited labeled data, which is advantageous for distinguishing open classes using confidence thresholds. However, as these models are trained with more labeled data, their performance tends to deteriorate. This is because, with the aid of the powerful feature extraction capabilities of deep neural networks, they have a tendency to produce high-confidence predictions even for open intent samples [14].

Moreover, it has been observed that OpenMax and DeepUnk are two competitive benchmarks. This is likely due to their ability to exploit the characteristics of intent feature distribution for open class detection. OpenMax calculates the centroids of each known class using only positive training samples for correction, but the centroids are easily affected by the number of training samples. On the other hand, DeepUnk employs a density-based novelty detection algorithm for open classification, which is also constrained by prior knowledge. As a result, both methods experience a significant drop in performance with fewer labeled data, as demonstrated in Fig. 7.4.

7.5.4 *Effect of Known Classes*

Varying ratios of known classes (25%, 50%, and 75%) were tested, and the findings are presented in Tables 7.2 and 7.3. In Table 7.2, we observed that ADB outperformed all other methods across all three datasets, displaying significant improvements. As the number of known classes decreased, all baselines exhibited a sharp decline in performance. In contrast, our method maintained robust accuracy scores even with limited training samples.

Upon examining Table 7.3, we can observe the meticulous performance of each method. It is apparent that all of the baseline models achieve high scores when identifying known classes, yet they encounter limitations in identifying open intent and exhibit suboptimal performance when presented with open classes. Nonetheless, our proposed approach yields the most favorable outcomes for both known and open classes. This reinforces the notion that our model's learned decision boundaries are beneficial for achieving a balanced open classification performance, regardless of the class being known or open. To summarize, these results underscore the importance of suitable decision boundaries in achieving optimal classification performance for both known and open classes.

7.6 Conclusion

In this section, we introduce a new technique for post-processing open intent classification. Our approach involves pre-training the model with annotated samples, which allows the model to learn precise and tailored decision boundaries that are adaptable to the given intent feature space. Notably, our method does not necessitate any alterations to the open intent or model architecture. Our comprehensive experiments on three standard datasets demonstrate that our method produces considerable enhancements compared to the competing baselines, and it exhibits greater robustness with fewer annotated data and a smaller number of known intents.

Summary of this part: Unknown intent detection is a critical and challenging research problem in dialogue systems. This part focuses on using deep neural networks' powerful feature extraction capabilities to discover new intents better. First, a new intent detection method SMDN based on model post-processing, is introduced to make neural network classifiers capable of detecting new intents. Any neural network classifier can detect new intents without modifying the model architecture. The method is divided into two parts. The first one is to detect new intentions through the probability threshold. The Softmax activation function is studied, and the sample confidence output of the classifier is corrected by temperature scaling to obtain a more reasonable probability distribution and threshold. The second is to combine deep intent representation with density-based anomaly detection algorithms for novel intent detection. Finally, combine the above two parts for joint prediction. The outcomes of the experiments indicate that this approach possesses robust generalization capabilities and is applicable to both single-turn and multi-turn dialogues. Additionally, this technique can enhance the efficacy of new intent detection tasks significantly.

Secondly, this chapter further introduces a large marginal cosine loss function in the neural network classifier to replace the Softmax cross-entropy loss function, forcing the model to consider the marginal item penalty while learning the decision boundary and minimizing the intra-class variance while maximizing the inter-class variance. Therefore, inter-class separation and intra-class compact intent representation can be obtained to facilitate the detection of new intents. Experimental results show that new intentions can be detected more easily. The experimental results show that the method introduced in this chapter surpasses all baseline algorithms and achieves the current best performance.

Since the static threshold as a decision boundary may introduce more open space risks, this article also learns a compact and adaptive decision boundary for each category of intent features. It effectively identifies unknown intentions by defining the outside of the decision boundary as an open space area.

References

1. Lin, T.-E., Xu, H., Zhang, H.: Discovering New intents via constrained deep adaptive clustering with cluster refinement. Proceedings of the 24th Association for the Advancement of Artificial Intelligence Conference on Artificial Intelligence, pp. 8360–8367 (2020)
2. Fei, G., Liu, B.: Breaking the closed world assumption in text classification. Proceedings of the 14th Conference of the North American Chapter of the Association for Computational Linguistics: Human Language Technologies, pp. 506–514 (2016)
3. Schölkopf, B., Platt, J.C., Shawe-Taylor, J., et al.: Estimating the support of a high-dimensional distribution. Neural Comput. **13**(7), 1443–1471 (2001)
4. Scheirer, W.J., De Rezende, R.A., Sapkota, A., et al.: Toward open set recognition. Inst. Electr. Electron. Eng. Transact. Pattern Analys. Mach. Intellig. **35**(7), 1757–1772 (2012)
5. Tapaswi, M., Law, M.T., Fidler, S.: Video face clustering with unknown number of clusters. Proceedings of the 15th Institute of Electrical and Electronics Engineers International Conference on Computer Vision, pp. 5026–5035 (2019)
6. Bendale, A., Boult, T.: Towards open world recognition. Proceedings of the 38th Institute of Electrical and Electronics Engineers Conference on Computer Vision and Pattern Recognition, pp. 1893–1902 (2015)
7. Casanueva, I., Temcinas, T., Gerz, D., et al. Efficient intent detection with dual sentence encoders. Proceedings of the 2nd Workshop on Natural Language Processing for Conversational Artificial Intelligence, pp. 38–45 (2020)
8. Larson, S., Mahendran, A., Peper, J.J., Clarke, C., Lee, A., Hill, P., Kummerfeld, J.K., Leach, K., Laurenzano, M.A., Tang, L., Mars, J.: An evaluation dataset for intent classification and out-of-scope prediction. Proceedings of the 58th Conference on Empirical Methods in Natural Language Processing and International Joint Conference on Natural Language Processing, pp. 1311–1316 (2019)
9. Xu, J., Wang, P., Tian, G., et al.: Short text clustering via convolutional neural networks. Proceedings of the 1st Workshop on Vector Space Modeling for Natural Language Processing, pp. 62–69 (2015)
10. Bendale, A., Boult, T.E.: Towards open set deep networks. Proceedings of the 39th Institute of Electrical an Electronics Engineers Conference on Computer Vision and Pattern Recognition, pp. 1563–1572 (2016)
11. Hendrycks, D., Gimpel, K.: A baseline for detecting misclassified and out-of-distribution examples in neural networks. Proceeding of the 5th International Conference on Learning Representations (2017)
12. Shu, L., Xu, H., Liu, B.: DOC: Deep open classification of text documents. Proceedings of the 22nd Conference on Empirical Methods in Natural Language Processing, pp. 2911–2916 (2017)
13. Lin T.N., Xu, H.: Deep unknown intent detection with margin loss. Proceedings of the 57th Annual Meeting of the Association for Computational Linguistics, pp. 5491–5496 (2019)
14. Nguyen, A., Yosinski, J., Clune, J.: Deep neural networks are easily fooled: high confidence predictions for unrecognizable images. Proceedings of the 38th Institute of Electrical and Electronics Engineers conference on computer vision and pattern recognition, pp. 427–436 (2015)

Part IV
Discovery of Unknown Intents

Preface This part introduces two methods to tackle the challenging new intents discovery problem. Having successfully separated unknown intents from known intents, this article is more concerned with what types of new intents have been discovered. Since most of dialogue data lacks annotation, this research makes an exploratory attempt to find new intents through the clustering algorithm. However, under the guidance of lack of prior knowledge, it is difficult for unsupervised clustering algorithm to obtain ideal clustering results because the same set of data may have a variety of different cluster division methods. Therefore, this research defines new intent discovery as a semi-supervised clustering problem and proposes two strong self-supervised constrained clustering methods.

Chapter 8
Discovering New Intents Via Constrained Deep Adaptive Clustering with Cluster Refinement

Abstract Identifying new user intents is an essential task in the dialogue system. However, it is hard to get satisfying clustering results since the definition of intents is strongly guided by prior knowledge. Existing methods incorporate prior knowledge by intensive feature engineering, which not only leads to overfitting but also makes it sensitive to the number of clusters. In this chapter, we introduce constrained deep adaptive clustering with cluster refinement (CDAC+), an end-to-end clustering method that can naturally incorporate pairwise constraints as prior knowledge to guide the clustering process. Moreover, the clusters are refined by forcing the model to learn from the high confidence assignments. After eliminating low confidence assignments, the approach presented is surprisingly insensitive to the number of clusters. Experimental results on the three benchmark datasets show that the method presented can yield significant improvements over strong baselines.

Keywords Dialogue system · Prior knowledge · Feature engineering · Deep adaptive clustering · Pairwise constraints

8.1 Introduction

In prat III, we have successfully separated the unknown type of new intent from the known intent through the new intent detection method. Of more concern, however, is exactly what new intents have been discovered. Since most dialogue data are not labeled, effective clustering methods can help us automatically find a reasonable classification system. But the problem is not so simple. First, the exact number of new intents is hard to estimate. Secondly, the intent taxonomy is usually defined according to human experience and has a variety of classification criteria. It is difficult to obtain ideal clustering results under the guidance of lack of prior knowledge. Finally, existing algorithms not only require a lot of feature engineering, but also carry out intent representation learning and cluster center allocation in a pipeline way, resulting in poor clustering performance and poor generalization of the model.

To solve these problems, this book introduces a new intent-discovery model CDAC+ based on self-supervised constrained clustering, an end-to-end deep clustering algorithm. First, the CDAC+ algorithm can jointly optimize the intent representation and cluster center allocation in the clustering process, eliminating the tedious feature engineering. Additionally, it leverages a limited quantity of labeled information as a reference point to steer the clustering procedure, substantially enhancing the efficacy of clustering. Finally, it eliminates cluster center allocation with low confidence through cluster refining module, making the algorithm insensitive to the number of cluster centers.

The rest of this chapter is arranged as follows. First, a new intent discovery model based on self-supervised constrained clustering is introduced in Sect. 8.2. Secondly, the experimental results and analysis are introduced in Sect. 8.3, and the performance of the proposed algorithm under different settings is discussed, including the number of different clustering centers, the proportion of labeled data, the proportion of new categories and the unbalanced data. Finally, Sect. 8.3 is the summary of this chapter.

8.2 New Intent Discovery Model Based on Self-Supervised Constrained Clustering

The proposed method is divided into three steps: intent representation, pairwise-classification, and cluster refinement. The model architecture is shown in Fig. 8.1. Steps 1 and 2 repeatedly carried out until the upper and lower boundaries intersect. Afterward, step 3 is performed to enhance the clustering outcomes. For optimal viewing of the illustration, it is recommended to observe it in color. The network parameters that are static are denoted by blue blocks.

8.2.1 Intent Representation

To begin with, the intent representations are derived by utilizing the pre-trained BERT language model. For the ith sentence x_i within the corpus, all token embeddings $[C, T_1, \cdots, T_N] \in R^{(N + 1) \times H}$ are extracted from the ultimate hidden layer of BERT and conduct mean-pooling to obtain the average representation $e_i \in R^H$:

$$e_i = mean - pooling([C, T_1, \cdots, T_N]) \tag{8.1}$$

formula for computing intent representations $I_i \in R^k$ can be expressed as follows: given a sequence of length N and a hidden layer size of H, e_i is inputted into the clustering layer g to obtain the intent representation:

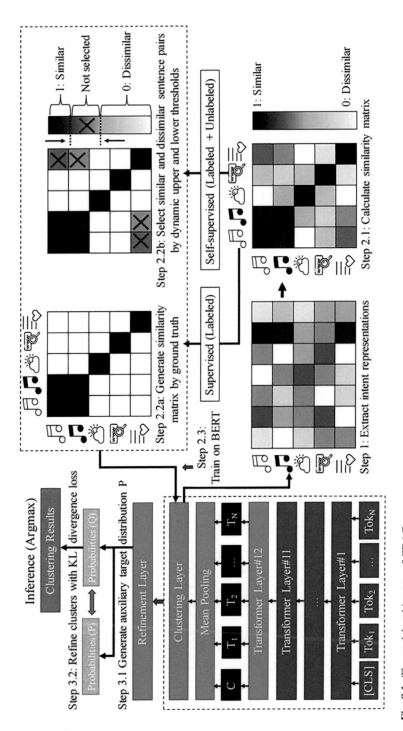

Fig. 8.1 The model architecture of CDAC+

$$g(e_i) = I_i = W_2(Dropout(tanh(W_1 e_i))) \tag{8.2}$$

where $W_1 \in R^{H \times H}$ and $W_2 \in R^{H \times k}$ are learnable parameters, and k is the number of clusters. The clustering layer is utilized to group the high-level features and extract intent representation I_i for the next steps.

8.2.2 Pairwise-Classification with Similarity Loss

The essence of clustering is to measure the similarity between samples [1, 2]. Motivated by DAC [3], we reinterpret the clustering challenge as a binary classification task between sentence pairs. Through discerning the similarity of each pair, the model presented can acquire an intent representation that facilitates clustering. The intent representation I is employed to calculate the similarity matrix S.

$$S_{ij} = \frac{I_i I_j^T}{\left\| I_i \right\| \left\| I_j \right\|} \tag{8.3}$$

Assuming n represents the batch size, S_{ij} denotes the similarity between sentence x_i and x_j, where i and j belong to the set $\{1, \cdots, n\}$. The model is optimized through a combination of supervised and self-supervised steps performed iteratively. The L2 norm, represented by $\| \cdot \|$, is utilized in the process.

Supervised Step

With a limited quantity of annotated data, it is possible to form a matrix R consisting of the labels.

$$R_{ij} = \begin{cases} 1, \text{if} & y_i = y_j \\ 0, \text{if} & y_i \neq y_j \end{cases} \tag{8.4}$$

The similarity loss, denoted as $\mathcal{L}_{sim}$, can be calculated by utilizing the similarity matrix S and the label matrix R. Here, it is assume that i and j are both integers ranging from 1 to n:

$$\mathcal{L}_{sim}(R_{ij}, S_{ij}) = -R_{ij} \log(S_{ij}) - (1 - R_{ij}) \log(1 - S_{ij}) \tag{8.5}$$

Labeled data as preexisting knowledge to direct the clustering procedure, indicating how the model should divide the data. This implies that the clustering process is guided by the prior information provided by the labeled data.

Unsupervised Step

Initially, through the implementation of varying thresholds on the similarity matrix S, the resultant self-labeled matrix $\widehat{R}$ is obtained.

$$\widehat{R}_{ij} := \begin{cases} 1, \text{if} S_{ij} > u(\lambda) \text{ or } y_i = y_j \\ 0, \text{if} S_{ij} < l(\lambda) \text{ or } y_i \neq y_j \\ Not\ selected, otherwise \end{cases} \tag{8.6}$$

The sentence pairs, denoted by i and j, are subject to similarity or dissimilarity determination using dynamic upper threshold $u(\lambda)$ and dynamic lower threshold $l(\lambda)$, where i and j are elements of $\{1, \cdots, n\}$. It should be noted that sentence pairs falling between $u(\lambda)$ and $l(\lambda)$ in terms of similarity are excluded from the training process. Labeled and unlabeled data are combined in this step to train the model. The incorporation of labeled data allows for the provision of ground truth, which can help to reduce errors in the noisy self-labeled matrix $\widehat{S}$.

Next, $u(\lambda) - l(\lambda)$ as a penalty term for the number of samples into the model.

$$\min E(\lambda) = u(\lambda) - l(\lambda) \tag{8.7}$$

The value of λ, which is an adaptive parameter regulating sample selection, is updated iteratively by us. This updating process follows the formula:

$$\lambda := \lambda - \eta \cdot \frac{\partial E(\lambda)}{\partial \lambda} \tag{8.8}$$

The λ can be gradually during training to decrease $u(\lambda)$ and increase $l(\lambda)$, where η represents the learning rate of λ. This approach allows for the gradual inclusion of more sentence pairs in the training process gradually. However, it may introduce more noise to $\widehat{S}$. This is because $u(\lambda)$ is proportional to $-\lambda$, while $l(\lambda)$ is proportional to λ. By gradually increasing λ, the balance between $u(\lambda)$ and $l(\lambda)$ can be adjusted.

The similarity matrix S and the self-labeled matrix can be calculated the similarity loss $\widehat{L}_{sim}$:

$$\widehat{L}_{sim}\left(\widehat{R}_{ij}, S_{ij}\right) = -\widehat{R}_{ij} \log\left(S_{ij}\right) - \left(1 - \widehat{R}_{ij}\right) \log\left(1 - S_{ij}\right) \tag{8.9}$$

with pairs of sentences that are progressively harder to classify as the thresholds are adjusted, in order to obtain a representation that is conducive to clustering. Once $u(\lambda) \leq l(\lambda)$, the iterative process is halted, and the refinement stage is initiated.

8.2.3 Cluster Refinement with KLD Loss

Concept presented in [4] and enhance the cluster assignments are enhanced using an iterative expectation maximization approach. The underlying idea is to facilitate the model in learning from the more certain assignments. To begin with, we utilize the initialized cluster centroids $U \in R^{k \times k}$ stored in the refinement layer and compute the soft assignment between intent representations and cluster centroids. The Student's t-distribution as a kernel function to measure the resemblance between the intent representation I_i and the centroid U_j of the cluster.

$$Q_{ij} = \frac{\left(1 + \|I_i - U_j\|^2\right)^{-1}}{\sum_{j'}\left(1 + \|I_i - U_j\|^2\right)^{-1}} \tag{8.10}$$

The variable Q_{ij} denotes the likelihood (soft assignment) that the instance i is a member of cluster j. Secondly, the auxiliary target distribution P is applied to compel the model to learn from the strong likelihood assignments, thus refining the parameters of the model and the cluster centroids. The target distribution P is defined as follows:

$$P_{ij} = \frac{Q_{ij}^2/f_i}{\sum_{j'} Q_{ij'}^2/f_{j'}} \tag{8.11}$$

where $f_i = \mathrm{sum}_i Q_{ij}$ denotes the soft cluster frequencies. Finally, the KLD loss between P and Q is minimized..

$$\mathcal{L}_{\text{KLD}} = KL\left(P\|Q\right) = \sum_i \sum_j P_{ij} \log \frac{P_{ij}}{Q_{ij}} \tag{8.12}$$

Then, the above two steps are repeated until the cluster assignment changes less than $\delta_{lable}\%$ in two consecutive iterations. Finally, we inference cluster c_i results as follows.

$$c_i = \operatorname*{argmax}_k Q_{ik} \tag{8.13}$$

where c_i is the cluster assignment for the sentence x_i.

8.3 Experiments

8.3.1 Datasets

Trials are carried out on three publicly accessible datasets containing brief texts. The comprehensive statistics are demonstrated in Table 8.1, where "#" represents the total count of sentences. For every experiment iteration, 25% of the intents at random are randomly designated as unknown. For instance, when considering the SNIPS dataset, 2 intents are randomly designated as unknown while acknowledging the remaining 5 intents as known.

1. **SNIPS**: The SNIPS dataset is comprised of 14,484 spoken phrases and encompasses 7 different categories of purposes for which they were spoken.
2. **DBPedia**: It comprises 14 distinct categories of ontology, which were chosen from DBPedia 2015 [5, 6]. As per the methodology outlined in [7], a set of 1000 examples were randomly selected for each of the aforementioned categories.
3. **StackOverflow**: Initially published on Kaggle.com, comprises 3,370,528 technical question titles that fall under 20 distinct categories. For this study, the dataset prepared by [8] was utilized, who employed random sampling to select 1,000 instances for each category.

8.3.2 Baseline

We compare the method proposed with both unsupervised and semi-supervised clustering methods.

1. **Unsupervised**: We compare the method proposed with K-means (KM) [9], agglomerative clustering (AG) [10], SAE-KM and DEC [11], DCN [12] and DAC [3]. To obtain embeddings for KM and AG, we utilize pre-trained GloVe [13] word embeddings and compute a 300-dimensional embedding by taking the average. Additionally, K-means clustering is applied to sentences that have been encoded with the average embeddings of all output tokens from the final hidden layer of a pre-trained BERT model (BERT-KM).
2. **Semi-unsupervised**: In evaluating semi-unsupervised approaches, comparisons were conducted with PCK-means [14], BERT-Semi [15], and BERT-KCL [16]. To ensure equitable comparisons, the backbone network of these methods was replaced with the same BERT model utilized in the approach.

Table 8.1 Statistics of SNIPS, DBPedia, and StackOverflflow dataset

Dataset	#Classes(Known + Unknown)	#Training	#Validation	#Test	Vocabulary	Length (max/mean)
SNIPS	7(5 + 2)	13,084	700	700	11,971	35/9.03
DBPedia	14(11 + 3)	12,600	700	700	45,077	54/29.97
StackOverflow	20(15 + 5)	18,000	1000	1000	17,182	41/9.18

8.3.3 Experiment Settings

Evaluation Metrics

Following previous studies and choose three metrics that are widely used to evaluate clustering results: Normalized Mutual Information (NMI), Adjusted Rand Index (ARI), and Clustering Accuracy (ACC). To calculate clustering accuracy, the Hungarian algorithm [17] is used to find the best alignment between the predicted cluster label and the ground-truth label. All metrics range from 0 to 1. The higher the score, the better the clustering performance.

Hyper Parameter

The model presented is constructed on the basis of the pre-existing BERT model (base-uncased variant, with 12-layer transformer) which has been implemented using PyTorch [18]. Most of the hyperparameter settings of the pre-trained model have been adopted. To prevent overfitting and accelerate the training process, all the parameters of BERT except the final transformer layer have been frozen. The training batch size is set to 256 and a learning rate of $5e^{-5}$. The dynamic thresholds are determined following the approach of DAC [3], where $u(\lambda) = 0.95 - \lambda$, $l-(\lambda) = 0.455 + 0.1 \cdot \lambda$, and $\eta = 0.009$.

In the refinement phase, K-means clustering is applied to the intent representation I to derive the initial centroids U, while the stopping criterion δ_{lable} is set to 0.1%.

8.3.4 Experiment Results

In each round of experimentation, a random selection process is used to designate 25% of classes as unknown and 10% of training data as labeled. The ground-truth number of clusters is used for our experiments and partition the entire dataset into three sets: training, validation, and test. Initially, the model is trained using limited labeled data (consisting of known intents) and unlabeled data (containing all intents) from the training set. Next, the model is fine-tuned using the validation set, which exclusively contains known intents. Finally, the model's performance is assessed on the test set. Ten rounds of experiments are conducted for each algorithm and report the average performance while ensuring no repetition of information.

The results are shown in Table 8.2. Both unsupervised and semi-supervised methods are evaluated. The introduces CDAC+ method outperforms other baselines by a significant margin in all datasets and evaluation metrics. It shows that the method presented effectively groups sentences based on the intent representations learned with pairwise classification and constraints, and even can generalize to new intents that are unknown in advance.

Table 8.2 The clustering results on three datasets

		SNIPS			DBPedia			StackOverflow		
	Method	NMI	ARI	ACC	NMI	ARI	ACC	NMI	ARI	ACC
Unsup.	KM	71.42	67.62	84.36	67.26	49.93	61.00	8.24	1.46	13.55
	AG	71.03	58.52	75.54	65.63	43.92	56.07	10.62	2.12	14.66
	SAE-KM	78.24	74.66	87.88	59.70	31.72	50.29	32.62	17.07	34.44
	DEC	84.62	82.32	91.59	53.36	29.43	39.60	10.88	3.76	13.09
	DCN	58.64	42.81	57.45	54.54	32.31	47.48	31.09	15.45	34.26
	DAC	79.97	69.17	76.29	75.37	56.30	63.96	14.71	2.76	16.30
	BERT-KM	52.11	43.73	70.29	60.87	26.6	36.14	12.98	0.51	13.9
Semi-sup.	PCK-means	74.85	71.87	86.92	79.76	71.27	83.11	17.26	5.35	24.16
	BERT-KCL	75.16	61.90	63.88	83.16	61.03	60.62	8.84	7.81	13.94
	BERT-semi	75.95	69.08	78.00	86.35	72.49	75.31	65.07	47.48	65.28
	CDAC+	**89.3**	**86.82**	**93,63**	**94.74**	**89.41**	**91.66**	**69.84**	**52.59**	**73.48**

The performance of unsupervised methods is particularly poor on DBPedia and StackOverflow, which may be related to the number of intents and the difficulty of the dataset. Semi-supervised methods are not necessarily better than unsupervised methods. If the constraints are not used correctly, it can not only lead to overfitting but also fail to group new intents into clusters.

Among these baselines, BERT-KM performed the worst, even worse than running K-means on sentences encoded with Glove. Our results suggest that fine-tuning is necessary for BERT to perform downstream tasks. Next, we will discuss the robustness and effectiveness of the proposed method from different aspects.

Ablation Study

To investigate the contribution of constraints and cluster refinement, we compare CDAC+ with its variant methods, such as performing K-means clustering with representation learned by DAC (DAC-KM) or CDAC (CDAC-KM), CDAC+ without constraints (DAC+) and CDAC+ without cluster refinement (CDAC). The results are shown in Table 8.3.

The addition of constraints has been shown to improve the performance of various methods. For example, CDAC+ has been found to significantly increase clustering accuracy compared to DAC+ on StackOverflow. This highlights the effectiveness of using constraints. When it comes to cluster refinement, both DAC + and CDAC+ consistently outperform DAC-KM and CDAC-KM. In fact, DAC+ even outperforms other baselines on SNIPS and DBPedia. This suggests that relying solely on learning representations through DAC or CDAC is insufficient, and that cluster refinement is necessary to achieve better results.

Table 8.3 The clustering results of CDAC+ and its variant methods

	Method	SNIPS			DBPedia			StackOverflow		
		NMI	ARI	ACC	NMI	ARI	ACC	NMI	ARI	ACC
Unsup.	DAC	79.97	69.17	76.29	75.37	56.30	63.96	14.71	2.76	16.30
	DAC-KM	86.29	82.58	91.27	84.79	74.46	82.14	20.28	7.09	23.69
	DAC+	86.90	83.15	91.41	86.03	75.99	82.88	20.26	7.10	23.69
Semisup.	CDAC	77.57	67.35	74.93	80.04	61.69	69.01	29.69	8.00	23.97
	CDAC-KM	87.96	85.11	93.03	93.42	87.55	89.77	67.71	45,65	71.49
	CDAC+	**89.30**	**86.82**	**93.63**	**94.74**	**89.41**	**91.66**	**69.84**	**52.59**	**73.48**

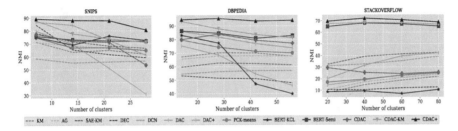

Fig. 8.2 Influence of the number of clusters on three datasets

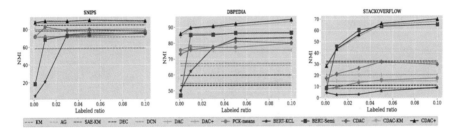

Fig. 8.3 Confusion matrix for the clustering results of CDAC+ on SNIPS datasets

Effect of the Number of Clusters

To study whether the proposed method is sensitive to the number of clusters or not, the number of predefined clusters is increased from its ground truth number to four times of it. The results are shown in Fig. 8.2. As the number of clusters increases, the performance of almost all methods except CDAC+ drops dramatically. Besides, the proposed method consistently performs better than CDAC-KM, which demonstrates the robustness of cluster refinement. In Fig. 8.3, the confusion matrix is used to analyze the results further. It shows that the proposed method not only maintains excellent performance, but also is insensitive to the number of clusters.

Effect of Labeled Data

In the experiments, the ratio of labeled data in the training set between 0.001 and 0.1, and plotted the results in Fig. 8.4. Notably, even when the labeled data ratio was as low as 0.001, the method presented CDAC+ outperformed most of the baselines. Additionally, it was observed that the performance changes significantly for the StackOverflow dataset, which has a taxonomy that can be categorized by technical subjects or question types (such as "what," "how," and "why"). This taxonomy requires prior knowledge in the form of labeled data to guide the clustering process. As a result, unsupervised methods fail to perform well, as there is no prior knowledge available to guide the clustering process. To summarize, our results demonstrate that CDAC+ is effective at achieving our goal of deduplication, especially when dealing with datasets that require prior knowledge to guide the clustering process.

Effect of Unknown Data

We vary the ratio of unknown classes in training set in the range of 0.25, 0.5 and 0.75, and show the results in Fig. 8.5. The higher the ratio of unknown classes, the more new intent classes in the training set exist. The method presented is still robust compared with baselines. In this case, the performance of BERT-Semi drops dramatically. The instance-level constraints they use will cause over-fitting and will not be able to group new intents into clusters.

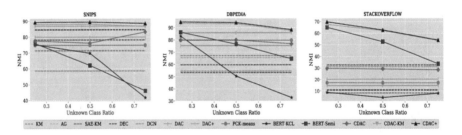

Fig. 8.4 Influence of the labeled ratio on three datasets

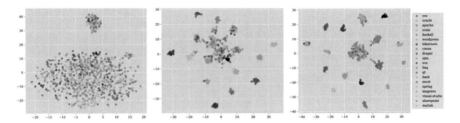

Fig. 8.5 Influence of the unknown class ratio on three datasets

Performance on Imbalanced Dataset

We follow previous works [3] and randomly sample subsets of datasets with different minimum retention probability γ. Given a dataset with N-classes, samples of class 1 will be kept with probability γ and class N with probability 1. The lower the γ, the more imbalanced the dataset is. The results are shown in Fig. 8.5. The method presented not only is robust to imbalanced classes, but also outperforms other baselines trained with balanced classes. The performance of other baselines drops around 3% to 10% under different γ.

Error Analysis

In the further, whether CDAC+ can discover new intents on the test set or not is analyzed deeply. BookRestaurant and SearchCreativeWork were designated as unknown entities in the training set illustrated in Fig. 8.3. The predefined number of clusters is twice of its ground truth. The values along the diagonal represent how many samples are correctly classified into the corresponding class. The larger the number, the deeper the color. Empty clusters are hidden for better visualization.

This method is still able to find out these intents. Note that some samples of SearchCreativeWork are incorrectly assigned to the cluster of SearchScrrenEvent, since they are semantically similar.

Using t-SNE [19], the intent representation is visualized in Fig. 8.6 and observed that the CDAC+ method yields a compact representation within each class and a distinguishable separation between classes. This suggests that the method presented effectively learns representations that promote clustering.

8.4   Conclusion

In this chapter, new intent discovery is defined as a semi-supervised clustering problem, and an end-to-end clustering model CDAC+ based on deep neural network is introduced. First, a small number of labeled samples and BERT pre-trained language model are regarded as prior knowledge to guide the clustering process. Secondly, a small number of labeled samples were converted into pairwise constraints, and a self-supervised similar binary task was constructed to learn clustering-friendly deep intent representation. Finally, the sample allocation with low confidence is eliminated by cluster refining, and the model is forced to learn from the high confidence allocation to further improve the clustering results.

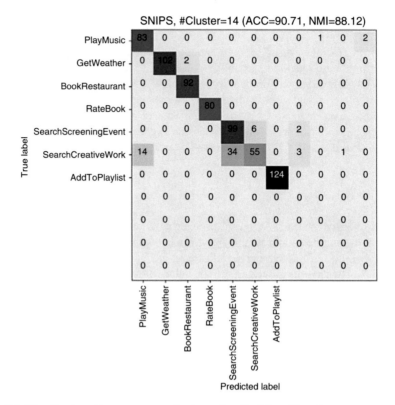

Fig. 8.6 Visualization of intent representation learned on StackOverflflow dataset

References

1. Hinton, G., Vinyals, O., Dean, J.: Distilling the knowledge in a neural network, arXiv 1503.02531 (2015)
2. Poddar, L., Neves, L., Brendel, W., et al.: Train one get one free: partially supervised neural network for bug report duplicate detection and clustering. Proceedings of the 17th Conference of the North American Chapter of the Association for Computational Linguistics: Human Language Technologies, pp. 157–165 (2019)
3. Chang, J., Wang, L., Meng, G., et al.: Deep adaptive image clustering. Proceedings of the 26th Institute of Electrical and Electronics Engineers International Conference on Computer Vision, pp. 5879–5887 (2017)
4. Liang, S., Li, Y., Srikant, R.: Enhancing the reliability of out-of-distribution image detection in neural networks. Proceedings of the 6th International Conference on Learning Representations (2018)
5. Wang, Z., Mi, H., Ittycheriah, A.: Semi-supervised clustering for short text via deep representation learning. Proceedings of the 20th Special Interest Group on Natural Language Learning Conference on Computational Natural Language Learning, pp. 31–39 (2016)
6. Zhang, X., Lecun, Y.: Text understanding from scratch. CoRR abs/1502.01710 (2015)
7. Lehmann, J., Isele, R., Jakob, M., et al.: DBpedia-a large-scale, multilingual knowledge base extracted from Wikipedia. Semant. Web. 6(2), 167–195 (2015)

8. Xu, J., Wang, P., Tian, G., et al.: Short text clustering via convolutional neural networks. Proceedings of the 1st Workshop on Vector Space Modeling for Natural Language Processing, pp. 62–69 (2015)
9. MacQueen, J., et al.: Some methods for classification and analysis of multivariate observations. Proceedings of the 5th Berkeley Symposium on Mathematical Statistics and Probability, pp. 281–297 (1967)
10. Gowda, K.C., Krishna, G.: Agglomerative clustering using the concept of mutual nearest neighbourhood. Pattern Recogn. **10**(2), 105–112 (1978)
11. Xie, J., Girshick, R., Farhadi, A.: Unsupervised deep embedding for clustering analysis. Proceedings of the 33rd International Conference on Machine Learning, pp. 478–487 (2016)
12. Yang, B., Fu, X., Sidiropoulos, N.D., et al.: Towards k-means-friendly spaces: simultaneous deep learning and clustering. Proceedings of the 34th International Conference on Machine Learning, pp. 3861–3870 (2017)
13. Pennington, J., Socher, R., Manning, C.: Glove: Global vectors for word representation. Proceedings of the 19th Conference on Empirical Methods in Natural Language Processing, pp. 1532–1543 (2014)
14. Basu, S., Banerjee, A., Mooney, R.J.: Active semi-supervision for pairwise constrained clustering. Proceedings of the 6th Society for Industrial and Applied Mathematics International Conference on Data Mining, pp. 333–344 (2004)
15. Bilenko, M., Basu, S., Mooney, R.J.: Integrating constraints and metric learning in semi-supervised clustering. Proceedings of the 21st International Conference on Machine learning, p. 11 (2004)
16. Hsu, Y.C., Lv, Z., Kira, Z.: Learning to cluster in order to transfer across domains and tasks. Proceedings of the 6th International Conference on Learning Representations (2018)
17. Kuhn, H.W.: The Hungarian method for the assignment problem. Naval Res. Logist. Quart. **2**(1–2), 83–97 (1955)
18. Thomas, W., Lysandre, D., Victor, S., et al.: Transformers: state-of-the-art natural language processing. Proceedings of the 26th Conference on Empirical Methods in Natural Language Processing: System Demonstrations, pp. 38–45 (2020)
19. Wang, F., Xiang, X., Cheng, J., et al.: Normface: L2 hypersphere embedding for face verification. Proceedings of the 25th Association for Computing Machinery International Conference on Multimedia, pp. 1041–1049 (2017)

Chapter 9
Discovering New Intents with Deep Aligned Clustering

Abstract Discovering new intents is a crucial task in dialogue systems. Most existing methods are limited in transferring the prior knowledge from known intents to new intents. These methods also have difficulties in providing high-quality supervised signals to learn clustering-friendly features for grouping unlabeled intents. In this work, we introduce an effective method (Deep Aligned Clustering) to discover new intents with the aid of limited known intent data. Firstly, by leveragin a few labeled known intent samples as prior knowledge to pre-train the model. Then, k-means is performed to produce cluster assignments as pseudo-labels. Moreover, an alignment strategy is proposed to tackle the label inconsistency problem during clustering assignments. Finally, the intent representations are learned under the supervision of the aligned pseudo-labels. With an unknown number of new intents, the number of intent categories is predicted by eliminating low-confidence intent-wise clusters. Extensive experiments on two benchmark datasets show that the method presented is more robust and achieves substantial improvements over the state-of-the-art methods.

Keywords Discovering new intents · Dialogue systems · Deep aligned clustering · Pseudo-labels · Alignment strategy

9.1 Introduction

Intent discovery has garnered significant attention in recent years [1–3]. Many researchers consider it to be an unsupervised clustering problem and attempt to integrate certain weak supervised signals to guide the clustering process. However, all of these methods are incapable of utilizing the prior knowledge of established intents. These approaches presuppose that the unlabelled samples solely comprise newly unidentified intents. A more typical scenario is that some labeled data pertaining to established intents is available and the unlabelled data contains a mixture of both established and new intents.

There are two main difficulties in this task. On the one hand, it is challenging to effectively transfer the prior knowledge from known intents to new intents with limited labeled data. On the other hand, it is hard to construct high-quality supervised

H. Xu et al., *Intent Recognition for Human-Machine Interactions*, SpringerBriefs in Computer Science, https://doi.org/10.1007/978-981-99-3885-8_9

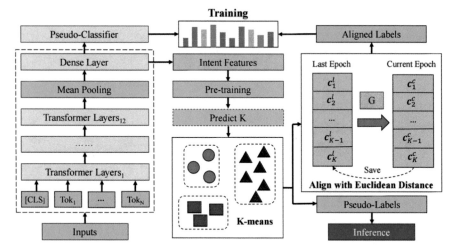

Fig. 9.1 The model architecture of the approach presented

signals to learn friendly representations for clustering both unlabeled known and new intents.

To solve these problems, we introduce an effective method to leverage the limited prior knowledge of intents and provide high-quality supervised signals for feature learning. As illustrated in Fig. 9.1, the approach firstly use the pre-trained BERT model [4] to extract deep intent features. Then, the model is pre-trained with the limited labeled data under the supervision of the Softmax loss. The pretrained parameters are retained and use the learning information to obtain well-initialized intent representations. Next, clustering is performed on the extracted intent features and estimate the cluster number K (unknown beforehand) by eliminating the low-confidence clusters.

As most of the training samples are unlabeled, we introduce an original alignment strategy to construct high-quality pseudo-labels as supervised signals for learning discriminative intent features. For each training epoch, k-means clustering is applied to the extracted intent features, and then use the produced cluster assignments as pseudo-labels for training the neural network. However, the inconsistent assigned labels cannot be directly used as supervised signals, the cluster centroids are utilized as the targets to obtain the alignment mapping between pseudo-labels in consequent epochs. Finally, another round of k-means is performed for inference. Benefit from the relatively consistent aligned targets, the method presented can inherit the history learning information and boost the clustering performance.

9.2 Deep Aligned Clustering

As shown in Fig. 9.1, the process begins by extracting intent representations with BERT. Then, knowledge transfer takes place from the limited labeled data of known intents. Finally, an alignment strategy is introduced to provide self-supervised signals for learning clustering-friendly representations.

9.2.1 Intent Representation

The pre-trained BERT model demonstrates its remarkable effect in NLP tasks [5], so it is used to extract deep intent representations. To begin with, the ith sentence, denoted as s_i, is inputted to BERT and extract all of its token embeddings $[CLS, T_1, \cdots, T_M] \in R^{(M+1) \times H}$ from the last hidden layer. Subsequently, mean-pooling is applied to generate the averaged feature representation of the sentence, which we refer to as $z_i \in R^H$:

$$z_i = mean - pooling([CLS, T_1, \cdots, T_M]) \qquad (9.1)$$

The vector CLS is utilized for text classification, where M represents the length of the sequence and H denotes the hidden size. To improve the feature extraction potential, we introduce a dense layer h and obtain the intent feature representation as $I_i \in R^D$:

$$I_i = h(z_i) = \sigma(W_h z_i + b_h) \qquad (9.2)$$

where σ is the Tanh, D represents the dimension of the intent representation, $W_h \in R^{M \times D}$ is the weight matrix, and $b_h \in R^D$ is the corresponding bias term.

9.2.2 Transferring Knowledge From Known Intents

To effectively transfer the knowledge, the limited labeled data is used to pre-train the model and leverage the well trained intent features to estimate the number of clusters for clustering.

Pre-training

The incorporate of limited prior knowledge is aimed at obtaining a good representation initialization for grouping both known and novel intents. As suggested in [6], the intent feature information is captured by pre-training the model with the labeled data. Specifically, feature representations are learned under the supervision of the cross-entropy loss. After pre-training, the classifier is removed and use the rest of the network as the feature extractor in the subsequent unsupervised clustering.

Predict K

In real scenarios, it is not always know in advance the number of new intent categories. In this case, it is necessary to determine the number of clusters K before clustering. Therefore, we introduce a simple and effective method to estimate K with the aid of the well-initialized intent features.

A big K' is initially assigned as the number of clusters (e.g., two times of the ground truth number of classes) at first. As a good feature initialization is helpful for partition-based methods (e.g., k-means) [4], utilizing the well pre-trained model to extract intent features. Then, k-means clustering is performed with the extracted features. It is suppose that real clusters tend to be dense even with K'. And the size of more confident clusters is larger than some threshold t. Low confidence clusters which size smaller t are discarded, and calculate K is calculated using the following approach.

$$K = \sum_{i=1}^{K'} \left(\delta |S_i| > = t \right) \tag{9.3}$$

where $|S_i|$ is the size of the ith produced cluster, and δ(condition) is an indicator function. It outputs 1 if condition is satisfied, and outputs 0 if not. Notably, the threshold t in this formula is determined by assigning it as the expected cluster mean size $\frac{N}{K'}$ in this formula.

9.2.3 Deep Aligned Clustering

After transferring knowledge from known intents, we introduce an effective clustering method to find unlabeled known classes and discover novel classes. We firstly perform clustering and obtain cluster assignments and centroids. Then, we introduce an original strategy to provide aligned targets for self-supervised learning.

Unsupervised Learning by Clustering

As most of the training data are unlabeled, it is important to effectively use a mass of unlabeled samples for discovering novel classes.

Inspired by DeepCluster [7], the proposed approach leverages the discriminative power of BERT to produce structured outputs as weak supervised signals. Initially, features are extracted from all training data using the pretrained model. Then, a standard clustering algorithm, k-means, is employed to learn both the optimal cluster centroid matrix C and assignments $\{y_i\}_{i=1}^{N}$:

$$\min_{c \in R^{M \times D}} \frac{1}{N} \sum_{i=1}^{N} \min_{y_i \in \{1, \cdots, K\}} \| I_i - C_{y_i} \|_2^2 \tag{9.4}$$

where N is the number of training samples and $\| \cdot \|_2^2$ denotes the squared Euclidean distance. The cluster assignments are subsequently utilized as pseudo-labels for feature learning.

Self-Supervised Learning with Aligned Pseudo-Labels

DeepCluster alternates between clustering and updating network parameters. It performs k-means to produce cluster assignments as pseudo-labels and uses them to train the neural network. However, the indices after k-means are permuted randomly in each training epoch, so the classifier parameters have to be re-initialized before each epoch [8]. Therefore, we introduce an alignment strategy to address the issue of inconsistent assignments.

We have observed that the centroid matrix C is not utilized in DeepCluster as shown in Eq. (9.4). However, C is a crucial part, which contains the optimal averaged assignment target of clustering. As each embedded sample is assigned to its nearest centroid in Euclidean space, the centroid matrix C is naturally employed as the prior knowledge to adjust the inconsistent cluster assignments in different training epochs. That is, this approach convert this problem into the centroid alignment. Though the intent representations are updated continually, similar intents are distributed in near locations. The centroid synthesizes all similar intent samples in its cluster, so it is more stable and suitable for guiding the alignment process. It is assumed that the centroids in contiguous training epochs are relatively consistently distributed in Euclidean space and adopt the Hungarian algorithm [9] to obtain the optimal mapping G.

$$C^c = G(C^l) \tag{9.5}$$

where C^c and C^l respectively denote the centroid matrix in the current and last training epoch. Then, aligned pseudo-labels y align are obtained using the mapping $G(\cdot)$.

$$y^{align} = G^{-1}(y^c) \tag{9.6}$$

where G^{-1} denotes the inverse mapping of G and y^c denotes the pseudo-labels in the current training epoch. Finally, the aligned pseudo-labels are used to perform self-supervised learning under the supervision of the softmax loss L_s.

$$L_s = -\frac{1}{N}\sum_{i=1}^{N} log \frac{exp\left(\varphi(I_i)^{y^{align}}\right)}{\sum_{j=1}^{K} exp\left(\varphi(I_i)^j\right)} \qquad (9.7)$$

where $\varphi(\cdot)$ is the pseudo-classifier for self-supervised learning, and $\varphi(\cdot)^j$ denotes the output logits of the jth class.

The cluster validity index (CVI) is used to evaluate the quality of clusters obtained during each training epoch after k-means. Specifically, the Silhouette Coefficient [10], an unsupervised metric, is adopted for evaluation.

$$SC = \frac{1}{N}\sum_{i=1}^{N} \frac{a(I_i) - b(I_i)}{max\{a(I_i), b(I_i)\}} \qquad (9.8)$$

The intra-class compactness of I_i, denoted as $a(I_i)$, is the average distance between I_i and all other samples in the ith cluster. The inter-class separation of I_i, denoted as $b(I_i)$, is the smallest distance between I_i and all samples not in the ith cluster. SC, the silhouette coefficient, ranges between -1 and 1, with a higher score indicating better clustering results.

9.3 Experiments

9.3.1 Datasets

In the experiments, two demanding benchmark intent datasets are used, and detailed statistics are provided in Table 9.1. The total number of sentences is indicated by "#". For each run of the experiment, 75% of the intents are randomly designated as known intents. To illustrate with the CLINC dataset, 113 intents are randomly chosen as known intents, while the remaining 37 intents are treated as new intents.

1. **CLINC**: The dataset [11] is an intent classification dataset that comprises 22,500 queries spanning across 10 domains and covering 150 intents.
2. **BANKING**: It is a fine-grained dataset in the banking domain [12], which contains 13,083 customer service queries with 77 intents.

9.3.2 Baselines

Unsupervised

In the comparison, unsupervised clustering methods are evaluated, including K-means (KM) [13], agglomerative clustering (AG) [14], SAE-KM, DEC [15], DCN [16], DAC [17], and DeepCluster [7].

Table 9.1 Statistics of CLINC and BANKING datasets

Dataset	#Classes (Known + unknown)	#Training	#Validation	#Test	Vocabulary	Length (max/mean)
CLINC	150 (113 + 37)	18,000	2250	2250	7283	28/8.31
BANKING	77 (58 + 19)	9003	1000	3080	5028	79/11.91

For KM and AG, the sentences are represented using the averaged pre-trained 300-dimensional word embeddings from GloVe [18]. For SAEKM, DEC, and DCN, the sentences are encoded using the stacked autoencoder (SAE), which is helpful to capture meaningful semantics on real-world datasets [15]. As DAC and DeepCluster are unsupervised clustering methods in computer vision, the backbone is replaced with the BERT model for extracting text features.

Semi-supervised

We also compare our method with semi-supervised clustering methods, including PCK-means [19], BERT-KCL [20], BERT-MCL [21], BERT-DTC [6] and CDAC+ [22]. For a fairness comparison, the backbone of these methods is replaced with the same BERT model as ours.

9.3.3   Experiment Settings

Evaluation Metrics

Three widely used metrics, namely Normalized Mutual Information (NMI), Adjusted Rand Index (ARI), and Accuracy (ACC), are adopted to evaluate the clustering results. The Hungarian algorithm is employed to calculate ACC by finding the optimal mapping between the predicted classes and the ground-truth classes.

Following the same settings as in [22], a 10% random subset of the training data is labeled, and 75% of all intents are designated as known. The datasets are split into training, validation, and test sets. The number of intent categories is set as ground-truth. Initially, the pre-training phase is conducted using a small labeled dataset consisting only of known intents. This pre-training process is further fine-tuned on the validation set. Then, self-supervised learning is performed using the entire training dataset, and the cluster performance is evaluated using the Silhouette Coefficient (as mentioned in (9.8)). Finally, the performance is assessed on the test set and report the averaged results over ten runs of experiments with different random seeds.

Hyper Parameters

The pre-trained BERT model (BERT-uncased, with 12-layer transformer) implemented in PyTorch [23] is used as our network backbone, and adopt most of its suggested hyper-parameters for optimization. The training batch size is 128, the learning rate is $5e-5$, and the dimension of intent features D is 768. Moreover, the last transformer layer parameters are the only ones not frozen to achieve better performance with BERT backbone, and speed up the training procedure as suggested in [22].

Experiment Results

The method presented consistently achieves the best results and outperforms other baselines by a large margin on all metrics and datasets, as shown in Table 9.2. The evaluation includes both unsupervised and semi-supervised clustering methods, with the best results highlighted in bold. This demonstrates the effectiveness of the method presented in discovering new intents with limited known intent data. Additionally, we observe that most semi-supervised methods perform better than unsupervised methods, indicating that even with limited labeled data as prior knowledge, it is helpful in improving the performance of unsupervised clustering.

Effect of the Alignment Strategy

The method presented is compared with the re-initializing strategy [7] to investigate the contribution of the alignment strategy. As shown in Table 9.3, the method presented has significant improvements over the re-initializing strategy on both semi-supervised and unsupervised settings. The reason is supposed to be that random initialization drops out the well trained parameters in the classifier in the former epochs. However, the method presented saves history embedding information by finding the mapping of pseudo-labels between contiguous epochs, which provides stronger supervised signals for representation learning.

Estimate K

To investigate the effectiveness to predict K, the experiment assigns K' as two times the ground truth number of classes and compares it with another two state-of-the-art

Table 9.2 The clustering results on two datasets

		CLINC			BANKING		
	Method	NMI	ARI	ACC	NMI	ARI	ACC
	KM	70.89	26.86	45.06	54.57	12.18	29.55
	AG	73.07	27.70	44.03	57.07	13.31	31.58
	SAE-KM	73.13	29.95	46.75	63.79	22.85	38.92
Unsupervised	DEC	74.83	27.46	46.89	67.78	27.21	41.29
	DCN	75.66	31.15	49.29	67.54	26.81	41.99
	DAC	78.40	40.49	55.94	47.35	14.24	27.41
	DeepCluster	65.58	19.11	35.70	41.77	8.95	20.69
	PCK-means	68.70	35.40	54.61	48.22	16.24	32.66
	BERT-KCL	86.82	58.79	68.86	75.21	46.72	60.15
Semi-supervised	BERT-MCL	87.72	59.92	69.66	75.68	47.43	61.14
	CDAC+	86.65	54.33	69.89	72.25	40.97	53.83
	BERT-DTC	90.54	65.02	74.15	76.55	44.70	56.51
	DeepAligned	**93.89**	**79.75**	**86.49**	**79.56**	**53.64**	**64.90**

Table 9.3 Effectiveness of the pre-training and the alignment strategy on two datasets

		CLINC			BANKING		
	Method	NMI	ARI	ACC	NMI	ARI	ACC
Without Pre-training	Reinitialization	57.80	9.63	23.02	34.34	4.49	13.67
	Alignment	62.53	14.10	28.63	36.91	5.23	15.42
With Pre-training	Reinitialization	82.90	45.67	55.80	68.12	31.56	41.32
	Alignment	**93.89**	**79.75**	**86.49**	**79.56**	**53.64**	**64.90**

Table 9.4 The results of predicting K with an unknown number of clusters

		CLINC ($K' = 300$)		BANKING ($K' = 154$)	
	Methods	K (Pred)	Error	K (Pred)	Error
25%	BERT-MCL	38	75.00	19	75.32
	BERT-DTC	94	37.33	37	51.95
	DeepAligned	**122**	**18.67**	**66**	**14.29**
50%	BERT-MCL	75	50.00	38	50.65
	BERT-DTC	**131**	**12.67**	**71**	**7.79**
	DeepAligned	130	13.33	64	16.88
75%	BERT-MCL	112	25.33	58	24.68
	BERT-DTC	195	30.00	110	42.86
	DeepAligned	**129**	**14.00**	**67**	**12.99**

methods (BERT-MCL and BERT-DTC). The ratio of known classes is varied in the range of 25%, 50%, and 75%, and calculate the error rate (the lower is better) for evaluation. As shown in Table 9.4, the method presented achieves the lowest or competitive results with different known class ratios. The ratio of known classes is varied in the range of 25%, 50% and 75%, and set K'as two times of the ground truth number of clusters during clustering. It shows the reasonability to estimate the cluster number by removing low-confidence clusters. We notice that BERT-DTC is a strong baseline, especially with 50% known classes. It relies on labeled samples to generate the probe set for determining the optimal number of classes. But the performance is unstable as the number of known intents changes. We also find the predicted K of BERT-MCL is close to the number of known classes, whose reason is that it uses known labeled data for classification while clustering. However, the classification information dominates in training and tends to predict known classes during testing.

Effect of Known Class Ratio

To investigate the influence of the number of known intents, the known class ratio is varied in the range of 25%, 50% and 75%. As shown in Figs. 9.2 and 9.3, the method presented still achieves the best results in all settings. All semi-supervised methods are vulnerable to the number of known intents. Particularly, though BERT-

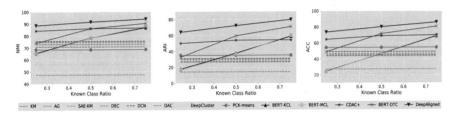

Fig. 9.2   Influence of the known class ratio on CLINC dataset

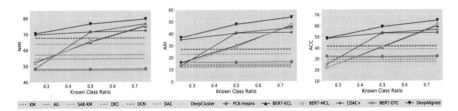

Fig. 9.3   Influence of the known class ratio on BANKING dataset

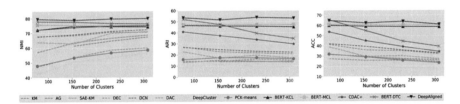

Fig. 9.4   Influence of the number of clusters on BANKING dataset

MCL and BERT-DTC are competitive with a 75% known class ratio, they drop dramatically as the known class ratio decreases. The reason for this could be that these methods largely depend on the prior knowledge of known intents to construct high-quality supervised signals (e.g., the pairwise similarity in BERT-MCL, initialized centroids in BERT-DTC) for clustering. In contrast, the method presented only needs known intent data for learning intent representations and is free from prior during the self-learning process. So the method presented performs more robust with fewer known classes.

Effect of the Number of Classes

To investigate the sensitiveness to the assigned cluster number K', The value of K' is varied from the ground-truth number to four times of it with 75% known classes. As shown in Figs. 9.4 and 9.5, the method presented achieves robust results even with large assigned cluster numbers. It is supposed that the benefit arises from a relatively accurate estimated cluster number. Additionally, it is observed that BERT-MCL and

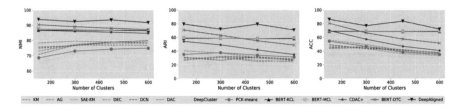

Fig. 9.5 Influence of the number of clusters on CLINC dataset

BERT-KCL seem to be insensitive to the number of classes. However, their performances largely depend on the prior knowledge of known intents.

9.4   Conclusion

In this work, we have introduced an effective method for discovering new intents. The method presented successfully transfers the prior knowledge of limited known intents and estimates the number of intents eliminating low-confidence clusters. Moreover, it provides more stable and concrete supervised signals to guide the clustering process. We conduct extensive experiments on two challenging benchmark datasets to evaluate the performance. The method presented achieves significant improvements over the compared methods and obtains more accurate estimated cluster numbers with limited prior knowledge. In the future, we will try different clustering methods besides k-means to produce supervised signals and explore more self-supervised learning methods.

Summary of this part: This part introduces the end-to-end self-supervised constrained clustering model CDAC+. Firstly, the clustering problem is transformed into a sentence pair similarity dichotomization problem to obtain cluster-friendly deep intent representation. Secondly, a small number of labeled samples were converted into pairwise constraints as prior knowledge to guide the clustering process, and the dynamic similarity threshold was used to conduct self-supervised learning. Finally, cluster center allocation with low confidence is eliminated by cluster refining, which greatly improves the robustness of the algorithm. The experimental results show that this method beats all the baseline methods of unsupervised and semi-supervised clustering, and maintains excellent performance under the condition that the cluster number, the proportion of labeled data, the proportion of known intents and the imbalanced category. Furthermore, we have introduced a highly effective approach for identifying novel intents. The approach presented successfully leverages prior knowledge from a limited set of known intents and accurately estimates the number of intents by eliminating low-confidence clusters. Additionally, it provides more robust and concrete supervised signals to guide the clustering process. We have conducted comprehensive experiments on two challenging benchmark datasets to evaluate the performance of the approach presented. The results show significant improvements compared to existing methods, and the approach presented yields more accurate estimates of cluster numbers with limited prior knowledge. Moving forward, we plan to explore alternative clustering techniques beyond k-means to generate supervised signals and investigate additional self-supervised learning methods.

References

1. Perkins, H., Yang, Y.: Dialog intent induction with deep multi-view clustering. Proceedings of the 58th Conferenceon Empirical Methods in Natural Language Processing, pp. 4016–4025 (2019)
2. Min, Q.K, Qin, L.B., Teng, Z.Y., et al.: Dialogue State Induction Using Neural Latent Variable Models. Proceedings of the 29th International Joint Conference on Artificial Intelligence, pp. 3845–3852 (2020)
3. Vedula, N., Lipka, N., Maneriker, P., Parthasarathy, S.: Open intent extraction from natural language interactions. Proceedings of the 29th Web Conference, pp. 2009–2020 (2020)
4. Platt, J.: Probabilistic outputs for support vector machines and comparisons to regularized likelihood methods. Adv. Large Marg. Classif. **10**(3), 61–74 (1999)
5. Devlin, J., Chang, M.-W., Lee, K., et al.: BERT: Pre-training of deep bidirectional transformers for language understanding. Proceedings of the 17th Conference of the North American Chapter of the Association for Computational Linguistics: Human Language Technologies, pp. 4171–4186 (2019)
6. Han, K., Vedaldi, A., Zisserman, A.: Learning to discover novel visual categories via deep transfer clustering. Proceedings of the 16th International Conference on Computer Vision, pp. 8400–8408 (2019)
7. Caron, M., Bojanowski, P., Joulin, A., et al.: Deep clustering for unsupervised learning of visual features. Proceedings of the 15th International Conference on Computer Vision, pp. 132–149 (2018)
8. Zhan, X., Xie, J., Liu, Z., et al.: Online deep clustering for unsupervised representation learning. Proceedings of the 43rd Institute of Electrical and Electronics Engineers Conference on Computer Vision and Pattern Recognition. pp. 6688–6697 (2020)
9. Kuhn, H.W.: The Hungarian method for the assignment problem. Naval Res. Logist. Quart. **2**(1–2), 83–97 (1955)
10. Rousseeuw, P.J.: Silhouettes: a graphical aid to the interpretation and validation of cluster analysis. J. Comput. Appl. Math. **20**, 53–65 (1987)
11. Larson, S., Mahendran, A., Peper, J.J., et al.: An evaluation dataset for intent classification and out-of-scope prediction. Proceedings of the 58th Conferenceon Empirical Methods in Natural Language Processing, pp. 1311–1316 (2019)
12. Casanueva, I., Temcinas, T., Gerz, D., et al.: Efficient intent detection with dual sentence encoders. Proceedings of the 2nd Workshop on Natural Language Processing for Conversational Artificial Intelligence, pp. 38–45 (2020)
13. MacQueen, J., et al. Some methods for classification and analysis of multivariate observations. Proceedings of the 5th Berkeley Symposium on Mathematical Statistics and Probability, pp. 281–297 (1967)
14. Gowda, K.C., Krishna, G.: Agglomerative clustering using the concept of mutual nearest neighbourhood. Pattern Recogn. **10**(2), 105–112 (1978)
15. Xie, J., Girshick, R., Farhadi, A.: Unsupervised deep embedding for clustering analysis. Proceedings of the 33rd International Conference on Machine Learning, pp. 478–487 (2016)
16. Schölkopf, B., Platt, J.C., Shawe-Taylor, J., et al.: Estimating the support of a high-dimensional distribution. Neural Comput. **13**(7), 1443–1471 (2001)

17. Chang, J., Wang, I., Meng, G., et al.: Deep adaptive image clustering. Proceedings of the 15th Institute of Electrical and Electronics Engineers International Conference on Computer Vision, pp. 5879–5887 (2017)
18. Pennington J, Socher R, Manning C.: Glove: Global vectors for word representation. Proceedings of the 19th Conference on Empirical Methods in Natural Language Processing, pp. 1532–1543 (2014)
19. Basu, S., Banerjee, A., Mooney, R.J.: Active semi-supervision for pairwise constrained clustering. Proceedings of the 6th Society for Industrial and Applied Mathematics International Conference on Data Mining, pp. 333–344 (2004)
20. Hsu, Y.C., Lv, Z., Kira, Z.: Learning to cluster in order to transfer across domains and tasks. Proceedings of the 6th International Conference on Learning Representations (2018)
21. Hsu, Y.-C., Lv, Z., Schlosser, J., Odom, P., Kira, Z.: Multi-class classification without multi-class labels. Proceedings of the 16th International Conference on Computer Vision (2019)
22. Lin, T.E., Xu, H.: Deep unknown intent detection with margin loss. Proceedings of the 57th Annual Meeting of the Association for Computational Linguistics, pp. 5491–5496 (2019)
23. Thomas, W., Lysandre, D., Victor, S., et al.: Transformers: state-of-the-art natural language processing. Proceedings of the 26th Conference on Empirical Methods in Natural Language Processing: System Demonstrations, Online. Association for Computational Linguistics, pp. 38–45 (2020)

Part V
Dialogue Intent Recognition Platform

Preface This part introduces a dialogue intent experiment platform: TEXTOIR*. TEXTOIR is the first open intent recognition platform to integrate and visualize textual data. It consists of two main functional modules, open intent detection and open intent discovery, which integrate the most advanced algorithms. In addition, this article illustrates the unified Open Intent Identification (OIR) framework in details, which connects two modules in a pipelined fashion to achieve a combination of multiple model functions. A range of convenient visualization tools for data and model management, training, and evaluation are provided by the platform. The performance of OIR is analyzed from various aspects, and the characteristics of different methods are presented. Finally, a comprehensive process of identifying known intents, discovering open intents, and recommending clustering keywords for open intents is implemented on a series of benchmark datasets.

Chapter 10
Experiment Platform for Dialogue Intent Recognition Based on Deep Learning

Abstract This chapter propose the first open intent recognition plat form TEXTOIR, which integrates two complete modules: open intent detection and open intent discovery. It provides toolkits for each module with common interfaces and integrates multiple advanced models and benchmark datasets for the convenience of further research. Additionally, it realizes a pipeline framework to combine the advantages of two modules. The overall framework achieves both identifying known intents and discovering open intents. A series of visualized surfaces help users to manage, train, evaluate, and analyze the performance of different methods.

Keywords Open intent recognition plat form · Open intent detection · Open intent discovery · Toolkits · Visualized surfaces

10.1 Introduction

In human-machine interaction services, such as dialogue systems, analyzing user intents is critical. However, current dialogue systems often struggle with recognizing user intents in uncertain open scenarios, as they are confined to closed world scenarios. The predefined categories may not be sufficient to cover all user requirements due to the varied and uncertain nature of user intents. As shown in Fig. 10.1, specific purposes like Flight Booking and Restaurant Reservation are easy to identify. However, there may be unrelated user utterances with open intents that need to be distinguished from known intents. Distinguishing these open intents from known intents can help improve service quality and uncover fine-grained classes for mining potential user requirements.

Open intent recognition (OIR) is divided into two modules: open intent detection and open intent discovery. The first module focuses on identifying known intents with n-class classification and detecting open intents with one-class classification [1–3]. It can identify known classes but fail to discover specific open classes. The second module further groups the one-class open intent into multiple fine-grained intent-wise clusters [4–6]. Despite this, the recognized categories cannot be identified by the employed clustering techniques.

© The Author(s), under exclusive license to Springer Nature Singapore Pte Ltd. 2023
H. Xu et al., *Intent Recognition for Human-Machine Interactions*, SpringerBriefs in Computer Science, https://doi.org/10.1007/978-981-99-3885-8_10

Fig. 10.1 An example for open intent recognition

User utterances	Intent Label
Book a flight from LA to Madrid.	Flight Booking
Can you get me a table at Steve's?	Restaurant reservation
Book Delta ticket Madison to Atlanta.	Flight Booking
Schedule me a table at Red Lobster.	Restaurant reservation
...	...
Can you tell me the name of this song?	**Open Intent$_1$**
What is the calorie of this food?	**Open Intent$_2$**

The two modules have achieved huge progresses with various advanced methods on benchmark datasets. However, there still exist some issues, which bring difficulties for future research. Firstly, there are no unified and extensible interfaces to integrate various algorithms for two modules, bringing challenges for further model development. Secondly, the current methods of the two modules are lack of convenient visualized tools for model management, training, evaluation and result analysis. Thirdly, the two modules both have some limitations for OIR. That is, neither of them can identify known intents and discover open intents simultaneously. Therefore, OIR remains at the theoretical level, and it needs an overall framework to connect the two modules for finishing the whole process.

We introduce TEXTOIR, the first platform for intent recognition in text that is integrated and visualized, as a solution to address the issues. The platform incorporates the following features.

The toolkit offers open intent detection and open intent discovery toolkits, respectively. It includes flexible interfaces for data, configuration, backbone, and method integration. The toolkit integrates advanced models for two modules, each supporting a complete workflow including data and backbone preparation with assigned parameters, training, and evaluation. It also provides standard and convenient modules for adding new methods. For more detailed information, please refer to https://github.com/thuiar/TEXTOIR.

By naturally combining two sub-modules, an overall framework is designed to achieve a complete Open Intent Recognition (OIR) process. The framework seamlessly integrates the strengths of both modules, allowing for automatic identification of known intents and discovery of open intent clusters using recommended keywords.

We offer a user-friendly interface that provides visualized surfaces for easy utilization. Users can take advantage of the methods provided or incorporate their own datasets and models for open intent recognition. Our platform includes front-end interfaces for two modules and a pipeline module. Each module supports model training, evaluation, and detailed result analysis using various methods. The pipeline module integrates both modules and displays comprehensive text OIR results. For more detailed information, please visit https://github.com/thuiar/TEXTOIR-DEMO.

10.2   Open Intent Recognition Platform

The architecture of the TEXTOIR platform is depicted in Fig. 10.2, which comprises four main modules. The initial module incorporates standard benchmark datasets. The second and third modules consist of toolkits for open intent discovery and open intent detection respectively. Additionally, the entire process, encompassing model management, training, evaluation, and result analysis of these two modules, is visually represented. Finally, the last module utilizes the two aforementioned modules in a pipeline framework to accomplish open intent recognition.

10.2.1   Data Management

The intent recognition module of our platform provides support for various standard benchmark datasets, including CLINC [7], BANKING [8], SNIPS [9], and StackOverflow [10]. These datasets are divided into training, evaluation, and test sets to facilitate model development and evaluation.

As depicted in Fig. 10.3, our unified data processing interfaces offer support for data preparation in two different formats. This includes sampling labeled data and known intents with assigned parameters for training and evaluation. In addition, the remaining unlabeled data are also utilized for open intent discovery. Detailed statistical information can be viewed by users on the frontend webpage, allowing them to manage their datasets efficiently.

10.2.2   Models

Our platform provides toolkits with standard and flexible interfaces, integrating a series of advanced and competitive models for two modules.

Open Intent Detection

For training, this module utilizes partial labeled known intent data to identify known intents and detect samples that do not belong to known intents. During testing, these detected samples are grouped into a single open intent class. The integrated methods are divided into two categories: threshold-based and geometric feature-based methods.

The methods based on thresholds consist of MSP [11], DOC [2], and OpenMax [12]. These techniques begin by being trained under supervision for the recognized intent classification task. They then utilize a probability threshold to detect open intent samples with low confidence. In contrast, the geometrical feature-based

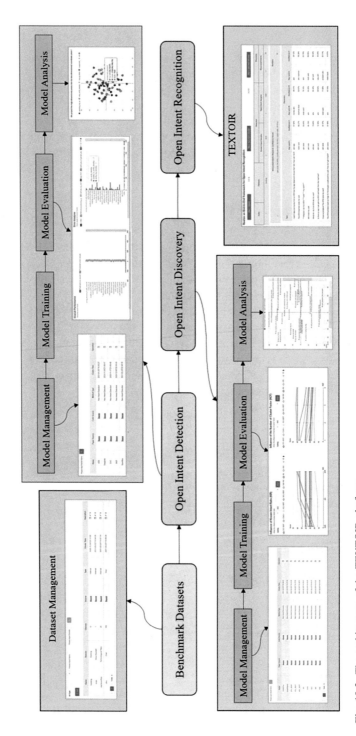

Fig. 10.2 The architecture of the TEXTOIR platform

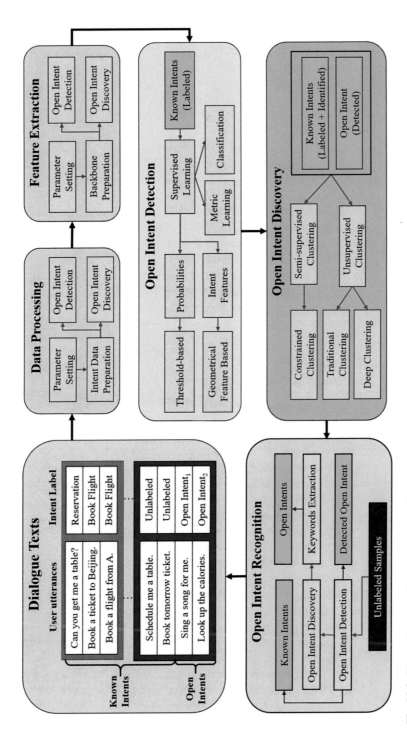

Fig. 10.3 The architecture of open intent recognition

methods include DeepUnk [9] and ADB [1]. DeepUnk utilizes the metric-learning approach to learn discriminative intent features, and the density-based methods to identify open intent samples as anomalies. Additionally, ADB leverages the boundary loss to learn adaptable decision boundaries.

Open Intent Discovery

This module uses both known and open intent samples as inputs, and aims to obtain intent-wise clusters by learning from similarity properties with clustering technologies. As suggested in [13] the integrated methods are divided into two parts, including unsupervised and semi-supervised methods.

The unsupervised methods include K-Means(KM) [14], agglomerative clustering (AG) [15], SAE-KM, DEC [16], and DCN [17]. The first two methods adopt the Glove [18] embedding, and the last three methods leverage stacked auto-encoder to extract representations. These methods do not need any labeled data as prior knowledge and learn structured semantic-similar knowledge from unlabeled data.

The semi-supervised methods include KCL [19], MCL [3], DTC [20], CDAC+ [4] and DeepAligned [13]. These methods can further leverage labeled known intent data for discovering fine-grained open intents.

10.2.3 Training and Evaluation

Our platform provides a visual interface for model training and evaluation, allowing users to easily adjust model hyperparameters and monitor the training process. Once training starts, the system automatically creates a record of the training process, and users can view and monitor changes in this status at any time. After successful completion of training, the platform also automatically saves the trained model and related parameters for further use.

In terms of model evaluation, we observe the prediction results from different perspectives to provide users with a comprehensive understanding of the model's performance. First, we display the number of correct and incorrect samples for each intent category to show the overall performance of the model. This helps users understand the model's performance on different intent categories and identify potential issues. Based on the overall performance, we further display the number of errors in fine-grained prediction categories to help users analyze which intents are prone to be confused by the model. This helps users gain a deeper understanding of the model's error patterns and provides targeted suggestions for improving the model. Additionally, line charts are used to show the impact of known intent proportion and label distribution on model performance. This helps users understand the characteristics of the dataset and the influence of label distribution on model performance, enabling them to better adjust their training strategy. Our platform also supports users in observing results on different selected datasets and evaluation metrics. This allows users to conduct multidimensional model evaluation and comparison according to their own needs and focus, helping them better understand the performance of the model in different contexts.

Through the visual display and analysis mentioned above, our platform helps users better understand the process of model training and evaluation, enabling them to more effectively optimize the model and improve its performance.

10.2.4   Result Analysis

Open Intent Detection

This module displays recognized samples of known intent and detected samples of open intent in the results, visualizing them using different confidence score thresholds to help users better understand the distribution of known and open intent at different probability thresholds. This visualization method can assist users in selecting appropriate probability thresholds to have better control over the output results of the model, as shown in Fig. 10.4.

Geometrical-based methods typically visualize intent representations on a two-dimensional plane by mapping high-dimensional intent representations to a lower-dimensional space. One commonly used dimensionality reduction technique is t-SNE (t-distributed Stochastic Neighbor Embedding), t-SNE can transform high-dimensional features into a lower-dimensional representation, making it convenient for visualization on a two-dimensional plane.

In the visualization, in addition to displaying the two-dimensional coordinates of intents, auxiliary information such as the center and radius of the Auxiliary Data Boundary (ADB) can also be shown for each point. This auxiliary information helps users better understand the distribution of intent features, enabling a more in-depth analysis and interpretation of the model's output results (Fig. 10.5).

Open Intent Discovery

In Fig. 10.6, we can observe the geometric positions and corresponding labels of the cluster centers generated by unsupervised and semi-supervised clustering methods. These centers are divided into two categories: known classes and open classes. Known classes are pre-defined categories, while open classes are newly generated classes during the clustering process.

In practical scenarios, due to the limitations of directly applying labels to clustering, the Key-BERT toolkit as a solution. By extracting keywords at both the sentence level and cluster level, we can obtain relevant information about open intents. Moreover, the toolkit is capable of calculating confidence scores for keywords in the cosine similarity space, further enhancing the accuracy of the keywords. In Fig. 10.7, we display the top three keywords for each discovered open intent along

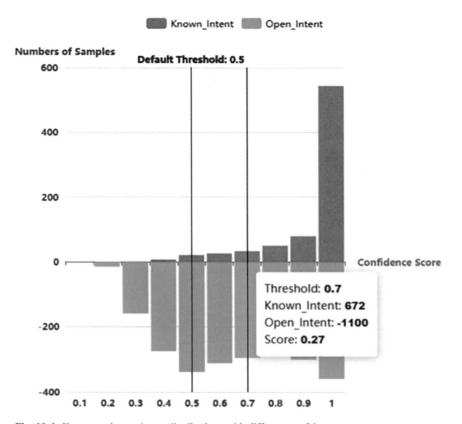

Fig. 10.4 Known and open intent distributions with different confidence scores

with their corresponding confidence scores, in order to gain a clearer understanding of the importance and credibility of the keywords.

10.3   Pipeline Framework

Open intent recognition and discovery are closely related natural language processing tasks aimed at identifying user intent from user input text. However, there is currently a lack of an integrated framework for effectively calling these two modules consecutively in a single process to achieve recognition of both known intents and open intents. To address this issue, TEXTOIR proposes a pipeline framework, as shown in Fig. 10.3.

The processing flow of the pipeline framework is as follows: Firstly, the original data is preprocessed and undergoes processing by two modules, including the labeling of known intent data and training of the open intent detection module. Then, unlabeled training data is predicted using a well-trained open intent detection

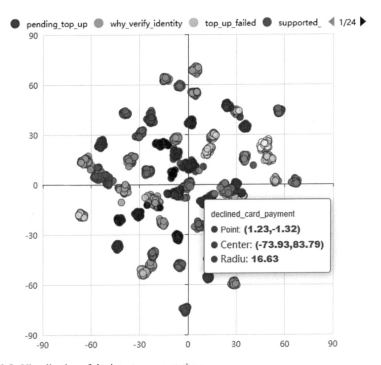

Fig. 10.5 Visualization of the intent representations

model to obtain more labeling information. Known intent data, detected open intents, and the original labeled data are identified as input for the open intent discovery module through evaluation results in the training data. Therefore, the discovery module can obtain richer input information from the detection module for further training. Next, the user's selected preferred clustering method is used to train the training data to obtain clustering results for open intents. Through this process, the pipeline framework can utilize information from known intents and detected open intents while processing the original data, thereby improving the discovery and clustering effectiveness of open intents.

When processing unannotated data for open intent recognition, a well-trained open intent detection method is initially utilized to predict known intents and detected open intents. Then, an open intent discovery method is employed to further predict the detected open intent data, resulting in more refined open intent clustering results. Finally, the KeyBERT toolkit is used as mentioned in Sect. 10.2.4 to extract keywords for each open intent cluster and add recommended labels based on these keywords. This way, our framework can better identify known intents and discover open intent samples through recommended labels with keywords, achieving more accurate and comprehensive open intent recognition tasks.

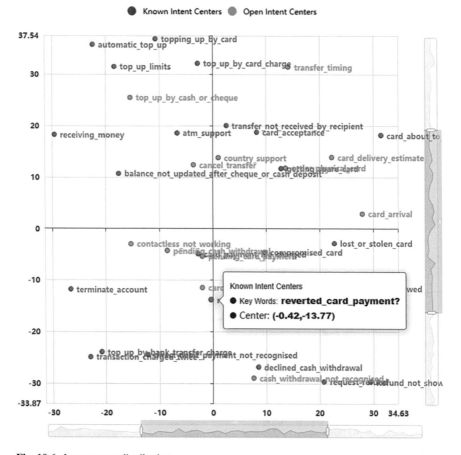

Fig. 10.6 Intent center distribution

10.4   Experiments

In this research, four intent benchmark datasets were used as mentioned in Sect. 10.2.1 to validate the performance of our TEXTOIR platform. In the experiments, different intent ratios, including 25%, 50%, and 75%, were set and tested under varying labeled ratio between 50% and 100%. The performance of the platform was evaluated by calculating the accuracy (ACC) on known intents and the normalized mutual information (NMI) on open intents. As components of the TEXTOIR platform, two advanced methods for open intent detection and discovery, namely ADB and DeepAligned, were employed. The experimental results are shown in Table 10.1, where "KIR" and "LR" represent known intent ratio and labeled ratio respectively, "Known" indicates the accuracy score on known intents, and "Open" represents the NMI score on open intents.

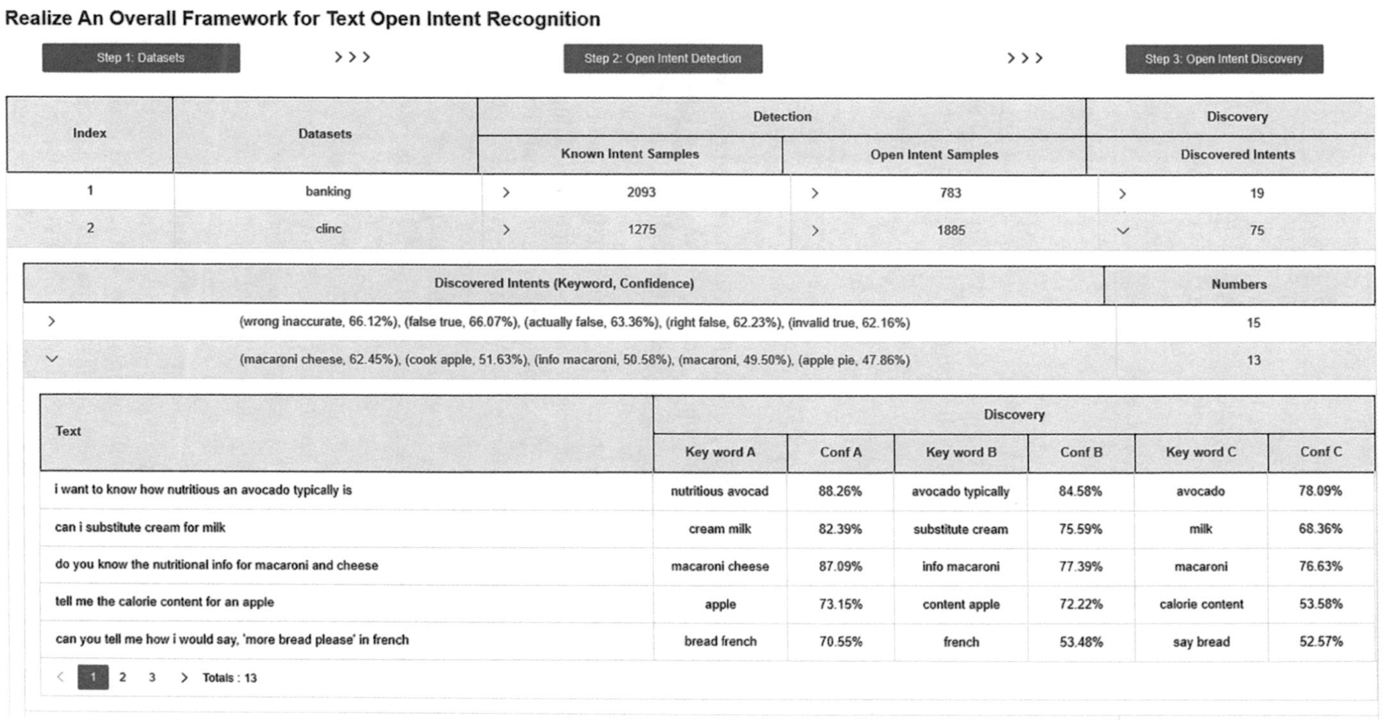

Realize An Overall Framework for Text Open Intent Recognition

Step 1: Datasets	› › ›	Step 2: Open Intent Detection	› › ›	Step 3: Open Intent Discovery

Index	Datasets	Detection		Discovery	
		Known Intent Samples	Open Intent Samples	Discovered Intents	
1	banking	∧ 2093	∧ 783	∧ 19	
2	clinc	∨ 1275	∨ 1885	∨ 75	

Discovered Intents (Keyword, Confidence)	Numbers
∧ (wrong inaccurate, 66.12%), (false true, 66.07%), (actually false, 63.36%), (right false, 62.23%), (invalid true, 62.16%)	15
∨ (macaroni cheese, 62.45%), (cook apple, 51.63%), (info macaroni, 50.58%), (macaroni, 49.50%), (apple pie, 47.86%)	13

Text	Discovery					
	Key word A	Conf A	Key word B	Conf B	Key word C	Conf C
i want to know how nutritious an avocado typically is	nutritious avocad	88.26%	avocado typically	84.58%	avocado	78.09%
can i substitute cream for milk	cream milk	82.39%	substitute cream	75.59%	milk	68.36%
do you know the nutritional info for macaroni and cheese	macaroni cheese	87.09%	info macaroni	77.39%	macaroni	76.63%
tell me the calorie content for an apple	apple	73.15%	content apple	72.22%	calorie content	53.58%
can you tell me how i would say, 'more bread please' in french	bread french	70.55%	french	53.48%	say bread	52.57%

‹ | 1 | 2 | 3 | › | Totals : 13

Fig. 10.7 The pipeline framework of open intent recognition

Table 10.1 The open intent recognition results of ADB + DeepAligned on four datasets

ADB + DeepAligned		CLINC		BANKING		SNIPS		StackOverflow	
KIR	LR	Known	Open	Known	Open	Known	Open	Known	Open
25%	50%	89.65	86.53	84.61	63.50	87.68	26.67	82.60	45.48
25%	100%	90.88	87.71	89.08	63.67	94.79	48.89	84.13	38.87
50%	50%	91.56	87.03	84.08	69.25	94.60	64.88	80.40	55.00
50%	100%	93.42	87.80	87.50	70.61	93.83	65.84	81.73	52.37
75%	50%	91.31	86.90	83.23	68.73	95.13	63.47	79.93	48.44
75%	100%	92.80	89.21	87.89	69.83	96.10	69.11	81.24	49.78

The pipeline framework has achieved remarkable success in connecting two modules, yielding competitive and robust results in various settings. Essentially, this framework addresses the shortcomings of the two modules and utilizes the first module to identify known intents, while leveraging the second module to discover unknown intents.

10.5   Conclusion

TEXTOIR is a leading intent recognition platform that integrates two complete modules, open intent detection and open intent discovery, in an innovative way. The platform provides toolkits for each module with standardized interfaces, incorporating advanced models and benchmark datasets, making it convenient for researchers to integrate and apply them. The platform utilizes an efficient pipeline framework that fully leverages the strengths of the two modules, achieving accurate recognition of known intents and discovery of unknown intents. The open intent detection module can identify known intents, while the open intent discovery module can uncover unknown intents, thereby improving the accuracy and robustness of intent recognition. The platform also offers an intuitive visual interface, allowing users to easily manage, train, evaluate, and analyze performance. This user-friendly design makes it convenient and efficient to use the platform, helping researchers and developers to explore and optimize intent recognition algorithms.

In conclusion, TEXTOIR is a leading intent recognition platform that provides powerful tools and support for researchers and developers in the field of intent recognition through innovative integration, efficient pipeline framework, and intuitive visual interface.

Summary of this part: This section presents the world's first-ever Intent Recognition Platform that combines two comprehensive modules: open intent detection and open intent discovery. The common interface was released during the implementation of the platform, and several high-level models were integrated for these two functional modules. The platform built in this discussion also supports a set of baseline intent data sets through automated processing tools. The visual component helps users analyze model results and characteristics of different approaches. Finally, combining the advantages of these two modules, an overall framework of open intention recognition is implemented.

References

1. Lin, T.E., Xu, H.: Deep unknown Intent Detection with Margin Loss. Proceedings of the 57th Annual Meeting of the Association for Computational Linguistics, pp. 5491–5496 (2019)
2. Shu, L., Xu, H., Liu, B.: DOC: Deep Open Classification of Text Documents. Proceedings of the 22nd Conference on Empirical Methods in Natural Language Processing, pp. 2911–2916 (2017)
3. Yan, G., Fan, L., Li, Q., et al.: Unknown intent detection using Gaussian mixture model with an application to zero-shot intent classification. Proceedings of the 58th Annual Meeting of the Association for Computational Linguistics, pp. 1050–1060 (2020)
4. Lin, T.-E., Xu, H., Zhang, H.: Discovering new intents via constrained deep adaptive clustering with cluster refinement. Proc. 24th Assoc. Advanc. Artific. Intellig. Conf. Artif. Intellig. **34**(5), 8360–8367 (2020)
5. Perkins, H., Yang, Y.: Dialog intent induction with deep multi-view clustering. Proceedings of the 24th Conference on Empirical Methods in Natural Language Processing, pp. 4016–4025 (2019)
6. Vedula, N., Lipka, N., Maneriker, P., Parthasarathy, S.: Open intent extraction from natural language interactions. Proceedings of the 29th Web Conference, pp. 2009–2020 (2020)
7. Larson, S., Mahendran, A., Peper, J.J., et al. : An evaluation dataset for intent Classification and out-of-scope prediction. Proceedings of 24th Conferenceon Empirical Methods in Natural Language Processing, pp. 1311–1316 (2019)
8. Basu, S., Banerjee, A., Mooney, R.J.: Active semi-supervision for pairwise constrained clustering. Proceedings of the 6th Society for Industrial and Applied Mathematics International Conference on Data Mining, pp. 333–344 (2004)
9. Zhang, H., Xu, H., Lin, T.E.: Deep open intent classification with adaptive decision boundary. Proceedings of the 25th Association for the Advancement of Artificial Intelligence Conference on Artificial Intelligence, pp. 14374–14382 (2021)
10. Xu, J., Wang, P., Tian, G., et al.: Short text clustering via convolutional neural networks. Proceedings of the 1st Workshop on Vector Space Modeling for Natural Language Processing, pp. 62–69 (2015)
11. Hendrycks, D., Gimpel, K.: A baseline for detecting misclassified and out-of-distribution examples in neural networks. Proceedings of the 5th International Conference on Learning Representations (2017)
12. Bendale, A., Boult, T.E.: Towards open set deep networks. Proceedings of the 39th Institute of Electrical an Electronics Engineers Conference on Computer Vision and Pattern Recognition, pp. 1563–1572 (2016)
13. Zhang, H., Xu, H., Lin, T.E., et al.: Discovering new intents with deep aligned clustering. Proceedings of the 25th Association for the Advancement of Artificial Intelligence Conference on Artificial Intelligence, pp. 14365–14373 (2021)
14. MacQueen, J., et al.: Some methods for classification and analysis of multivariate observations. Proceedings of the 5th Berkeley Symposium on Mathematical Statistics and Probability. pp. 281–297 (1967)
15. Gowda, K.C., Krishna, G.: Agglomerative clustering using the concept of mutual nearest neighbourhood. Pattern. Recogn. **10**(2), 105–112 (1978)
16. Xie, J., Girshick, R., Farhadi, A.: Unsupervised deep embedding for clustering analysis. Proceedings of the 33rd International Conference on Machine Learning, pp. 478–487 (2016)

17. Yang, B., Fu, X., Sidiropoulos, N.D., et al.: Towards k-means-friendly spaces: Simultaneous deep learning and clustering. Proceedings of the 34th International Conference on Machine Learning. pp. 3861–3870 (2017)
18. Pennington, J., Socher, R., Manning, C.: Glove: Global vectors for word representation. Proceedings of the 19th Conference on Empirical Methods in Natural Language Processing, pp. 1532–1543 (2014)
19. Hsu, Y.C., Lv, Z., Kira, Z.: Learning to cluster in order to transfer across domains and tasks. Proceedings of the 6th International Conference on Learning Representations. (2018)
20. Han, K., Vedaldi, A., Zisserman, A.: Learning to discover novel visual categories via deep transfer clustering. Proceedings of the 16th International Conference on Computer Vision, pp. 8400–8408 (2019)

Part VI
Summary and Future Work

Chapter 11
Summary

Abstract This book focuses on the study of conversation intent understanding in Artificial Intelligence. The research objective is to enhance machine understanding of human intents in the real world. With the increasing demand for human-computer interaction, intelligent services have become integrated into our daily lives. The book explores the core module of understanding user needs during conversation interactions. Accurately recognizing user intents contributes to improved service quality and commercial value. Traditional intent recognition methods are limited in closed-world settings and are unable to address the diversity and uncertainty of the real world. Therefore, the book proposes two novel research tasks, namely open intent detection and open intent discovery, to tackle the challenges of the open world. These tasks have significant commercial applications, and corresponding algorithms and frameworks are presented. Additionally, the book identifies areas for Future Work, highlighting the need for further advancements and improvements in intent recognition techniques. It suggests exploring novel approaches, leveraging new data sources, and refining existing models to better address the complexities of real-world scenarios. To facilitate research in this area, the book provides a platform called TEXTOIR, which supports the implementation of these tasks and serves as a foundation for future work.

Keywords Intent recognition · Human-computer interaction · Future work

The research presented in this book focuses on conversation intent understanding in Artificial Intelligence. In particular, the research goal is to make the machine better understand human intents in the real world. With the increasing demand for human-computer interaction, intelligent services gradually incorporate into our daily life. A typical example is an intelligent customer, which can be applied in catering, medical care, e-commerce, etc. Moreover, there are also emerging intelligent ChatBots or voice assistants, such as Siri, Cortana, Google Assistant, etc.

The core module during conversation interaction is to understand user needs. Accurately recognizing user intents helps us improve service quality and bring more business value. In natural language understanding, intent analysis even acts as the" brain" in the dialogue system and directly determines the quality of subsequent

© The Author(s), under exclusive license to Springer Nature Singapore Pte Ltd. 2023   147
H. Xu et al., *Intent Recognition for Human-Machine Interactions*, SpringerBriefs in
Computer Science, https://doi.org/10.1007/978-981-99-3885-8_11

decision-making and feedback. Though intent recognition is well-studied in the literature, but most traditional intent recognition methods are restricted to closed-world settings, assuming all intent categories are accessible. However, it is inapplicable in real-world scenarios as user needs are usually various and uncertain.

We have explored three different perspectives, including research tasks, algorithms and frameworks, and benchmark datasets. Among them, we have addressed the challenges of the open world by proposing two novel research tasks, namely open intent detection and open intent discovery. Specifically, the open intent detection task treats all irrelevant utterances as a category of open intents. The goal is to leverage the prior knowledge of known intents from the training set to identify known intents and detect unseen open intents in the test set. The significance of this task lies in the fact that in real-world dialogue scenarios, users may pose novel, unseen questions or express diverse intents. Therefore, accurately identifying and handling these open intents is crucial for the performance and user experience of intelligent dialogue systems. The final performance needs a tradeoff between known-class identification and open class detection. It requires no need for the label-intensive and time-consuming process of collecting labeled data from the open class. However, there are also some limitations in open intent detection. It overly depends on labeled known intents during training and suffers performance degradation with less annotated data. Besides, it fails to leverage a large amount of unlabeled data in real applications and cannot find fine-grained open intents. Thus, we introduce the open intent discovery task to solve these problems. In this task, the training data contains only a small amount (e.g., 10%) of labeled samples from known intents. The remaining data are all unlabeled, containing both known and open intents. Open intent discovery aims to fully use the unlabeled data with the guidance of limited labeled data during clustering and explore clusters of known and open intents. The quality of produced intent-wise clusters determines the final performance.

These tasks are of great commercial value and have broad applications. As introduced in open-world challenges, the conventional closed-world intent recognition paradigm will bring false-positive errors with unexpected utterances, which is harmful to human-computer interaction devices (e.g., dialogue systems). Open intent detection can improve the system's robustness and provide better services. Open intent discovery can further explore potential user needs for profile mining and personalized recommendation. In particular, Expedia and Meituan have applied the open intent discovery task in a real business scenario to discover user consumption intents and enhance recommendation.

A series of novel and effective algorithms are designed for open intent detection and discovery. We also build a convenient toolkit for the two tasks and propose a unified framework to achieve open intent recognition. Two state-of-the-art algorithms, and DA-ADB, are proposed for open intent detection. ADB is a simple yet effective method, aiming to automatically learn adaptive decision boundaries in the known intent feature space. The initial step involves utilizing the pre-trained language model to capture latent intent features and conduct supervised representation learning with known intent labels. Then, spherical decision boundaries are constructed for each known intent cluster. For each ball area, the centroid is computed by averaging all the embedded samples in the corresponding known class, and the radius is obtained by nonlinear mapping with learnable parameters. We have introduced a novel boundary loss that aims to balance empirical and open space risks in order to learn appropriate decision boundaries. The motivation behind this approach is to ensure that the boundaries are not only wide enough to encompass most of the known intent samples, but also close to centroids in order to avoid introducing open-intent samples. DAADB improves ADB by learning more discriminative representations with distance-aware concepts. It enables perceiving whether a sample is" easy" or" hard" by comparing the distance difference between the nearest and the next-nearest known-class centroids. The distance information is further utilized to calibrate initial intent representations during training. The two methods achieve significant improvements in average performance (2%–14% ACC) over strong baselines on three intent benchmark datasets. Moreover, they are more robust to different proportions of labeled data and known classes.

Two most popular algorithms, CDAC+ and DeepAligned, are proposed for open intent discovery. CDAC+ makes the first trial on this task, an end-to-end clustering method successfully transferring the knowledge from labeled data to learn semantic similarity relationships. The motivation is that labeled data can be used as guidance to tell whether a sentence pair is similar or not. Besides, it eliminates low-confidence cluster assignments by learning from an auxiliary target distribution. However, as the constructed pairwise similarities are weak supervised signals, CDAC+ is limited in distinguishing specific classes and performs worse on the challenging dataset with many similar intents. To address this issue, we introduce a deep aligned clustering (DeepAligned) algorithm for open intent discovery. One of the state-of-the-art methods is DeepAligned, which generates high-quality pseudo-labels for learning clustering-friendly representations by effectively transferring knowledge from limited labeled data. The model is first pre-trained with limited labeled data to provide a solid initialization for clustering. Then, we hope clustering and representation learning can benefit each other. To address the issue of label inconsistency caused by the randomness in initial cluster centroid selection, an original alignment strategy is introduced. However, the cluster assignments cannot be used directly as supervised signals due to their inner properties. The motivation is that the distributions of cluster centroids in Euclidean space are relatively consistent and stable between adjacent iterations. As the centroids are exactly the assignment targets during clustering, they are used to obtain alignment projection of successive cluster assignments. Empirically, we find the aligned targets can gradually converge as the training process goes on, which is beneficial to representation learning. Furthermore, a simple yet effective method is proposed to deal with unknown cluster number scenarios by eliminating low-confidence clusters with cluster-size lower than some threshold.

To foster research within the community, we offer a platform called TEXTOIR (Text Open In-Tent Recognition) along with a pipeline framework and toolkit. The toolkit encompasses two modules that cater to open intent detection and discovery, supporting end-to-end workflows from data preparation to model training and

Appendix

Symbol cross-reference table

ADB	Adaptive-decision-boundary
AG	Agglomerative clustering
BERT	Bidirectional encoder representation from transformers
BiLSTM	Bi-directional long short-term memory
BPTT	Back-propagation through time
CBOW	Continuous bag-of-words model
CDAC	Constrained deep adaptive clustering
CDAC+	Constrained deep adaptive clustering with cluster refinement
CNN	Convolutional neural network
CRF	Conditional random fields
DM	Dialogue management
DNN	Deep neural network
DOC	Deep open classification
ECE	Expected calibration error
ELMo	Embedding from language models
Glove	Global vector
GPT	Generative pre-training
GRU	Gate recurrent unit
ID	In domain
IPA	Intelligent personal assistant
KB	Knowledge base
LLR	Linear logistic regression
LMCL	Large margin cosine loss
LOF	Local outlier factor
LSTM	Long short-term memory
Masked LM	Masked language model
MSP	Maximum Softmax probability
NER	Named entity recognition

(continued)

© Tsinghua University Press 2023

H. Xu et al., *Intent Recognition for Human-Machine Interactions*, SpringerBriefs in Computer Science, https://doi.org/10.1007/978-981-99-3885-8

Symbol cross-reference table   (continued)

N-gram	N-gram grammar
NLG	Natural language generation
NLP	Natural language processing
NLU	Natural language understanding
NMI	Normalized mutual information
NN	Neural network
NNLM	Neural network language model
NSP	Next sentence prediction
OIR	Open intent recognition
OOD	Out of domain
RNN	Recurrent neural network
RNNLM	Cyclic neural network language model
Skip-gram	Continuous skip-gram model
SLU	Spoken language understanding
SVDD	Support vector data description
SVM	Support vector machine
TF-IDF	Term frequency–inverse document frequency
t-SNE	t-distributed stochastic neighbor embedding

Code link table

SofterMax and deep novelty detection	https://github.com/thuiar/SMDN
Adaptive-decision-boundary	https://github.com/thuiar/Adaptive-Decision-Boundary
Constrained deep adaptive clustering with cluster refinement (CDAC+)	https://github.com/thuiar/CDAC-plus
DeepAligned-clustering	https://github.com/thuiar/DeepAligned-Clustering
TEXTOIR toolkit	https://github.com/thuiar/TEXTOIR
TEXTOIR demo	https://github.com/thuiar/TEXTOIR-DEMO

Printed in the United States
by Baker & Taylor Publisher Services